ISBN: 979-8-9991705-0-7 (Softcover)

ISBN: 979-8-9991705-2-1 (Hardcover)

ISBN: 979-8-9991705-1-4 (eBook)

Library of Congress Control Number (LCCN): 2025911963

Ilustrations, designs, and book layout by: Walter Policelli

First Edition

For information, contact:

MNRD Publishing,

Pearland, Texas, 77581 USA

contact@mnrdpublishing.com

www.mnrdpublishing.com

A POCKET GUIDE TO
SCIENTIFIC
WRITING
and
PUBLISHING

MARTÍN A. NUÑEZ

MNRD
Publishing

CONTENTS

INTRODUCTION

Let's be clear: I'm not a gifted or exceptional writer. This book isn't about turning you into one either. It's about getting things done—finishing the manuscripts you're struggling with and navigating the publishing process effectively so you can publish it in the best possible journal.

I've spent over 20 years in academia, published hundreds of papers, and participated in the editorial and review process for many more. I'm still learning, but I've figured out strategies that work—especially for those of us who aren't natural writers or for whom English isn't a first language (or both, as in my case).

Writing in academia is rarely easy, yet it's one of the most crucial skills for a scientist. It often feels like just another step in the research process, but in reality, it can be the biggest roadblock. If you're reading this, you probably already know that.

This isn't a book about creative writing. It's not about making your writing beautiful. It's about making it **clear, precise, and publishable**. And for non-native English speakers like me, that can be an especially steep challenge. But I have also seen plenty of native speakers that have very similar struggles. Here we will focus on strategies that work.

A bit about my writing journey

When you first start exploring what it means to "do science," you quickly realize the complexity of each step in the process—coming up with ideas, securing funding, collecting and analysing data, and ultimately, writing and publishing the results.

As an undergraduate, I was struck by how **writing papers in English** —something I initially saw as a minor step in comparison with actually doing the research—turned out to be one of the biggest challenges. Like most researchers in Latin America, I didn't grow up speaking English, so writing a scientific manuscript felt overwhelming, even more so than developing research questions or collecting data.

When I became a PhD student in the U.S., my **personal goal** was to become a good and prolific writer. I knew it was possible since I met people like my advisor, Dan Simberloff, and researchers like Dave Richardson, and others who seemed to write papers effortlessly. I wanted to learn how they did it.

Back in Argentina, I had been told that the biggest obstacle to publishing was **writing in English**, a language that often feels impossible to fully master (there are more exceptions than rules!). And while it's true that writing in a non-native language adds an unfair barrier, after years of publishing, I've come to realize that **English itself is not the biggest challenge**. This is especially clear now with AI polishing our English better than ever.

What really matters is **how you structure a paper, frame your ideas, and fine-tune your manuscript to get it published**—skills that are just as critical whether you're writing in English, Spanish, or Chinese. Yet, these aspects are rarely taught.

I started writing papers more than 20 years ago, and this book is my way of sharing everything I've learned—not just with my students,

but with anyone who wants to improve their scientific writing. From this book you may **learn something, agree with some things and disagree with others** (I am expecting that in the AI section!), but this is all great. Each one of us needs to develop our own styles and see what works for each one of us. We are all different, have different experiences and work on different topics, but I know that thinking about writing and publishing is good for everyone, especially at early stages as a researcher, when publishing can make a bigger difference.

This is a **practical guide** to help you start, finish, and publish your manuscripts. Let's get started.

#01

REASONS TO PUBLISH PAPERS AND CHALLENGES TO GET PUBLISHED

Why do we write papers?

Whenever I give talks about publishing, I always ask the audience why they think we write papers. The answer, in 100% of cases, is some version of: "To share our knowledge with others." This is absolutely true and a great reason—science is based on the exchange of ideas and information, and published peer reviewed papers are the base of that. Needless to say, science is fundamental for human progress and for addressing global challenges, driving innovation, and improving our understanding of nature. So publishing is pretty important.

However, sharing knowledge is not the only reason we publish peer review papers. Scientific careers come with practical needs: securing jobs, earning promotions, and building strong CVs to continue working in research. Some PhD programs even have requirements for a certain number of publications. These professional milestones are important, and publishing plays a key role in achieving them.

That said, the primary goal should always be to create meaningful impact through our work. By keeping both aspects in mind— knowledge-sharing and career-building—we can make thoughtful decisions about where and how we publish, ensuring our research reaches the right audience while also supporting our professional growth.

Understanding why we publish is crucial when writing papers and selecting journals. My PhD advisor—a highly accomplished scientist with over 500 papers, countless awards, and membership in the U.S. National Academy of Sciences—once told me: "Now with the internet, journals don't matter anymore. If you do good science, people will eventually find it and use it." I fully agree with him on the knowledge-sharing aspect, but I don't completely share his view when it comes to career-building. Some journals are more prestigious, have faster turnaround times, or reach a broader audience. Where and how you publish still matters.

In this book, I take a very pragmatic approach to the entire process. My goal is to help you improve your writing and navigate publishing more successfully.

The writing challenges for scientists and why it is so important to overcome them

Scientific writing is fundamentally different from writing a novel or other literature genres, yet most of us receive little to no formal training in it before entering academia. Scientific writing has two primary goals: clarity (leaving no room for confusion) and precision (avoiding ambiguity and providing information at the required level of detail) (see glossary). While making the text flow well and sound elegant is certainly important, these aspects are somehow secondary. Papers with adequate methods and meaningful results are rarely rejected for overly long sentences or lack of literary elegance; they are rejected for fundamental flaws—in logic, clarity, or precision. Of course, a well-polished manuscript improves your chances of acceptance and can give your work a wider readership, but before polishing, your priority should be ensuring clarity and precision.

The challenges of writing and publishing are especially hard for early-career researchers and for non-native English speakers who do not have strong writing skills in English (a group that, in reality,

includes many people—even native speakers). Many young scientists struggle with structuring their arguments, knowing how much detail to include, or figuring out the subtle expectations of different journals. For non-native speakers, the added challenge is expressing complex ideas in a language they may not fully master, often leading to rejections not because of poor science but because of difficulties in communication. The problem is compounded by the pressure to publish in high-impact journals, where polished writing can make a significant difference in acceptance. Unfortunately, academia rarely provides formal training to bridge this gap, leaving researchers to learn through trial and error, feedback from reviewers, or the guidance of mentors—if they are lucky enough to have supportive ones.

I often feel that writing is an overvalued skill in science. You can have the ideas, do all the work, but if you cannot write it down, your contribution is significantly diminished. On the other hand, if someone has done little to develop the ideas or carry out the research but is skilled at writing, they may still find themselves in a privileged position in the author list. This imbalance highlights the power of scientific writing—not just as a tool for communication but as a gatekeeper for recognition and career advancement.

#02

COMMON BARRIERS FOR WRITING AND TO GET WRITING DONE

Identify your strengths and weaknesses as a writer

Imagine these two scenarios of researchers who struggle with writing:

1. **The Perfectionist Writer** – Your writing is excellent. Your grammar is flawless. Your paper structure is meticulously designed for clarity. But it takes forever to complete a draft, and sometimes you stop halfway, afraid to share incomplete work.

2. **The Fast but Messy Writer** – Your writing is full of typos. Your sentences are unclear and overly long. Your paper lacks structure, and the message is muddled. But you get words down quickly and are not shy about sharing imperfect drafts for feedback.

Surprisingly, the poor writer in this scenario has an advantage—they produce text. And you can't improve or publish what isn't written. This is something I see all the time: people who write inefficiently but consistently tend to publish more than those who aim for perfection and never finish.

Knowing your strengths and weaknesses as a writer is essential for improving your writing process.

Every researcher has different strengths. Some are great at creating figures that tell a compelling story, others excel at producing detailed drafts quickly, while some are naturals at structuring outlines or crafting engaging titles. Recognizing your strengths allows you to leverage them strategically—either to compensate for your weaknesses or to collaborate effectively (see Chapter 6: The Role of Collaborators).

At the same time, it's crucial to identify your weaknesses. Do you struggle with grammar, clarity, argument flow, or structuring sections? If so, don't try to fix everything alone. Seek help from colleagues, mentors, or writing tools that can assist with specific issues. The key is to be honest about where you need improvement and to find effective strategies or people to help you—rather than struggling in isolation. As we will see below, writer #2 has a key advantage solely because they are happy to look out for help and to find collaborators on what they struggle with.

The shitty draft: just write something—anything

One of the biggest barriers to writing is the fear of producing something bad. Many early-career researchers (ECRs) know exactly what a polished scientific paper should look like but feel completely incapable of producing anything close to that standard. This gap between expectation and reality creates frustration, anxiety, and

often, writer's block.

The solution? **Write a terrible first draft**. A bad draft is better than no draft because once something is on the page, you have material to work with. No one publishes their first version of a manuscript—it always goes through rounds of editing. The important thing is to get words down, no matter how messy they seem. You can fix bad writing; you can't fix a blank page.

Perfection is the Enemy of Good (and Done): Striving for perfection is often what keeps people from finishing a paper. The reality is that no manuscript is ever perfect—there is always something that could be tweaked, improved, or rewritten. But if you keep waiting for a perfect draft, you may never submit your work at all.

Instead of aiming at producing a perfect paper, focus on getting your paper to **"good enough"**—clear, well-structured, and scientifically sound. Journals have reviewers and editors for a reason: if they think that the writing problems are big enough, they will let you know. The goal isn't to write a flawless paper —it's to submit the results of your research to be shared with others (and to gain freedom to move into your next project!).

The sooner you embrace this mindset, the sooner you'll finish more papers, submit more work, and ultimately become a better writer.

Tips for better writing

1. Read what you wrote out loud

Reading your manuscript out loud helps you catch awkward phrasing, unclear sentences, and overly complex structures. If a sentence sounds strange when spoken, it likely needs to be rewritten for better clarity and flow.

2. Read books about writing

Books on scientific writing offer valuable insights and practical techniques to improve your writing. Many contain useful tips that can make your papers clearer, more persuasive, and easier to read. Writing is a skill, and like any skill, it improves with deliberate practice and guidance.

3. Read a lot of papers

The best way to understand how to write well is to read widely. Pay attention to structure, phrasing, and argumentation in high-quality papers. Identify what works and what doesn't, what you like and what you don't like—this will help you refine your own writing style.

4. Do not be afraid to talk to editors

Editors are not mythical gatekeepers; they are scientists like you. If you meet an editor at a conference or workshop, don't hesitate to introduce yourself and discuss your research. They appreciate knowing about interesting new work, and engaging with them can give you insights into what their journal is looking for. And you stop being an anonymous author next time you submit to that journal which can help.

5. Review papers—It's one of the best ways to learn

Reviewing papers puts you on the other side of the process, allowing you to see what makes a paper strong (or weak). Many journals also provide access to other reviewers' comments, which is incredibly educational. You'll learn what reviewers focus on, which will help you anticipate critiques of your own work. **How to Get Papers to Review? It's Easier Than You Think!** If you're an early-career researcher, the easiest way to start reviewing is by co-reviewing with a senior researcher (e.g., your advisor or a mentor). Many senior scientists receive more review requests than they can handle and will

gladly involve you in the process. Simply ask them if you can co- review a paper with them—they will often appreciate the help, and you will gain valuable experience. Once you have co-reviewed a few papers and published some, you may start getting your own papers to review.

#03
EDITING IS EASIER THAN CREATING

If you think about it, this is obvious: **editing is easier than creating**. If I ask you to write a page about your favorite experience last summer, you might stare at a blank screen, struggling to come up with anything worth sharing. But if I give you a rough draft on that topic and ask you to edit it, you will almost certainly find ways to improve it.

This is a central premise behind the **shitty first draft** approach—writing something, no matter how bad, is always better than staring at a blank page. Most people, even excellent scientists, struggle to produce polished text from scratch. However, many of those same people enjoy editing because it's **easier to improve something than to create from nothing.**

My PhD advisor, a prolific writer and master of the English language, once told me that he knew only one person who could write a near-final draft on the first attempt: **Stephen Jay Gould**. But let's be honest—most of us are not Stephen Jay Gould (if you don't know him, take a look at his inspiring work on evolution). The rest of us rely on **multiple rounds of editing** to refine our ideas and make our writing clear and compelling.

The power of 48-hour buddies

One of the best writing habits I developed in grad school was having **"48-hour buddies"**—fellow students willing to read each other's rough drafts. The emails typically started with something like:

"Could you please read this really raw, incomplete, typo-filled draft? I can't look at it for one more minute!"

You send the draft to your buddy, and they give it back with feedback within 48 hours. The key to this system is **a tacit agreement to be extra positive.** At this early stage, an idea is fragile

—it's easy to criticize, but the real value comes from helping each other refine the work into something a step closer to be publishable.

This system works incredibly well and has the bonus benefit of helping you build a **network of colleagues** who support each other's writing. Of course, once your draft is more polished (it will never be perfect!), you can send it to your **PI or other senior researchers** for feedback. Sending a **very early draft** to a PI may not be the best use of their time unless you need advice on the fundamental direction of the paper. Your peers, however, are great resources for catching clarity issues and fixing early structural problems.

If you're in the mood to write, keep writing

When you feel inspired to write, cancel unnecessary meetings, skip that seminar, and focus on writing. For some of us, writing motivation can be rare—when you find yourself in that state, take full advantage of it. Sometimes writing is the most important thing we can do, so feel free to prioritize it!

How to get writing done?

1. Sit down and wait for the urge to write

Sometimes, even simple tasks like doing laundry feel impossible. You know it needs to get done, but you just can't bring yourself to start. Writing is the same—whether you rely on inspiration or not, starting is always the hardest part.

The only way to get writing done is to **sit down and write**. Unfortunately, there's no shortcut for this. However, one major obstacle is that when people set aside four hours to write, they often

end up writing for only a fraction of those hours. The temptation to do something else—you name it: checking your plants in the greenhouse, running a new lab experiment, suddenly deciding that your office needs cleaning—is always strong.

One of the best strategies is to **stay at your desk and wait for the words to come**. If you need to take a break, **stay in front of your computer**—read an article, do some light editing, but do not go far from your writing space. If you resist the urge to escape, writing will eventually happen.

2. Writing groups: motivation and accountability

Many people find that writing in groups makes them more productive. This can be informal—just sitting with colleagues in the same room—or structured, where group members set **daily or monthly writing goals** and report on progress. Knowing that others are tracking your writing can serve as **extra motivation (or pressure!)** to stay on task. This works! Do it!

3. Writing with a timer

One of my favorite strategies is **writing in short, timed sessions**. I close my email, social media, and everything distracting, then set **a timer for 20 minutes**.

Twenty minutes may sound short, but it's much better than writing nothing. Most importantly, I can **force myself** to write for 20 minutes, no matter how much I don't feel like it. Often, something magical happens after that—I get into the rhythm of writing and can keep going for an hour or more. **It's like running—the first few minutes are the hardest, but once you get going, you can keep going.**

Deadlines, self-Imposed deadlines, and semi-self-imposed deadlines

Believe it or not, deadlines **help us get things done**. When the **National Science Foundation of the USA (NSF)** changed some of its grant programs from having two fixed submission deadlines per year to an open submission policy, what happened? **Submissions plummeted.** Researchers procrastinate without a hard deadline.

The same happens with writing. If your paper has no deadline, your brain tells you:

"Why do this now? Don't you have other things to finish first?"

Real deadlines—like conference submission deadlines, journal

revisions, or thesis defenses—are great motivators. But what about papers that have no formal deadlines?

One strategy I use is creating **semi-self-imposed deadlines**. The key is to make them feel real by **telling someone else** about them. If you're a PhD student, tell your advisor:

"I plan to send you a draft on this date. Does that work for you?"

By marking it on your calendar (and theirs), you create instant motivation (and pressure!). **The deadline must be realistic.** If your advisor is expecting your manuscript, you're much more likely to meet that deadline.

On "good reasons" to ignore your deadline

What if, by the deadline date, your manuscript still isn't ready? You may think, *"I just need two more months to make it perfect."*

Two things to keep in mind:

1.   **Your expectations for your own work may be too high.**

2.   **An imperfect manuscript turned in early is better than an imperfect manuscript turned in late.**

Often, after two months of additional work, a manuscript remains largely unchanged. Scientific writing involves multiple iterations, and **submitting an imperfect but functional draft keeps the process moving.** In some cases, significant issues spotted by others may require major revisions, making it inefficient to spend time deeply polishing a paper that isn't yet ready. Of course sometimes there are good reasons for waiting extra time and ignoring deadlines, such as if you are waiting for new data. However, in most cases sticking to the deadlines is the way to go.

Multiple iterations are fundamental

In an ideal world, you'd write your paper, get feedback from co-authors, incorporate their comments, submit, and get it published. **But this rarely happens.**

Papers evolve through **many iterations**. One of my PhD students once complained about working on **version #15** of a paper. But they

forgot that along the way, we had **changed the focus, added co-authors, and incorporated new data.** Writing is an iterative process, and feedback—even when painful—almost always makes a paper stronger.

The worst scenario is **too little feedback**. If your co-authors simply say, "Looks fine!" without engaging, it's often because they didn't read it carefully. While painful, **detailed comments mean people care enough to help improve your work.**

Embrace the process, expect multiple revisions, and keep writing!

#04

THE IMPORTANCE OF STRUCTURE

Many researchers think of writing as an abstract, creative process, but in reality, it can be a **mechanical exercise.** Structuring your paragraphs and sections properly makes writing easier, faster, more effective and even more enjoyable. If you know the basic framework, you can **focus on the content instead of struggling with organization.**

This chapter offers a practical, step-by-step approach to writing well-organized paragraphs and structuring an entire paper in a way that improves clarity, readability, and the chances of getting published.

Creative vs. mechanical writing

Writing is often seen as a purely creative process, but in academic publishing, a structured, mechanical approach can be far more effective. While creativity has its place—particularly in crafting an engaging title, designing figures, or developing clear explanations—relying too much on inspiration can slow progress and make writing inconsistent. Some people don't even like creative titles and prefer the purely descriptive ones! Understanding the mechanics of writing allows you to make progress even on days when you don't feel inspired.

A well-organized, structured approach ensures that writing becomes a repeatable process rather than an unpredictable creative effort. By mastering paper structure, paragraph organization, and systematic writing techniques, you can produce clear, publishable work efficiently, regardless of how "creative" you feel on a given day.

Creativity helps in scientific writing, but you don't necessarily need it

Creativity plays a role in academic writing by improving clarity, engagement, and impact. Titles benefit from creativity, as adding clarity or intrigue can make a paper more noticeable. Figures and visuals also rely on creative choices to present complex ideas in an accessible and engaging way.

Creativity is particularly valuable in introductions and discussions, where well-chosen analogies or compelling framing can enhance understanding. While results and methods follow a structured format, these sections require a more fluid narrative to highlight significance and broader implications. However, creativity should always serve clarity rather than being an end in itself. Titles should be engaging yet precise, figures should be visually appealing but scientifically accurate, and explanations should simplify without distorting meaning.

That said, a straightforward title or figure is perfectly acceptable, and many successful papers take this approach. Creativity is a useful tool, but not a requirement. If you are naturally creative, it can be an asset, but if you are not, clear and structured writing remains the priority. The writer that gets the writing done is always the winner. Constantly tweaking sentences or spending excessive time on figures can become forms of procrastination.

Focusing on mechanical writing: the power of structure and templates

One of the most effective ways to become a productive writer is to make the writing process as mechanical as possible. This means relying on clear, repeatable steps rather than waiting for inspiration. The more structured your approach, the easier it becomes to produce high-quality work consistently.

A scientific paper follows a predictable format, typically including a title, abstract, introduction, methods, results, and discussion. By treating each section as a set of predefined tasks rather than as a creative exercise, you can **break down the writing process into manageable steps.** The introduction moves from broad context to specific research questions. The methods describe what was done in a clear, replicable way. The results present findings without interpretation. The discussion interprets results, connects them to broader literature, and suggests future directions. See chapter 7 for more details on this.

Thinking of your paper as a structured set of components reduces the mental effort needed to start writing. Even if you feel uninspired, you can still make progress by completing structured sections one step at a time.

Papers have multiple levels of structure, from the overall organization of sections to the internal structure within each section and even down to individual paragraphs. A well-structured paper follows a logical progression, guiding the reader through the content smoothly and enhancing clarity. Maintaining coherence at all levels not only improves readability but also makes the writing process much more efficient and systematic.

Copy the structure of papers that you like: works like magic!

· E X E R C I S E ·

To practice writing structured paragraphs, try this exercise: Find a well-written paper in your field that is similar to the one you want to write. Analyze how paragraphs are structured by looking at the first sentence of each paragraph. Identify patterns—does the first sentence introduce a general idea? Does it frame the paragraph's content? Compare how different paragraphs develop their ideas and apply the same approach in your own writing. Rather than approaching each paragraph as a blank slate, this method helps you construct them systematically, making writing more mechanical and less dependent on inspiration.

Making writing mechanical to increase productivity

By treating writing as a structured process rather than a creative art, you can be productive even on days when writing feels difficult. Outlining before writing can help you avoid the blank-page problem and ensure that you always know what comes next. Using templates or following the structure of previously published papers in your field provides a clear model to follow. Setting small, achievable goals— such as writing a single paragraph rather than aiming to finish an entire section—keeps progress manageable and prevents writing from feeling overwhelming.

Breaking writing into short sessions can also make it easier to stay consistent. Writing in focused, 20–30-minute blocks prevents burnout and allows you to make steady progress without the pressure of needing to complete everything at once.

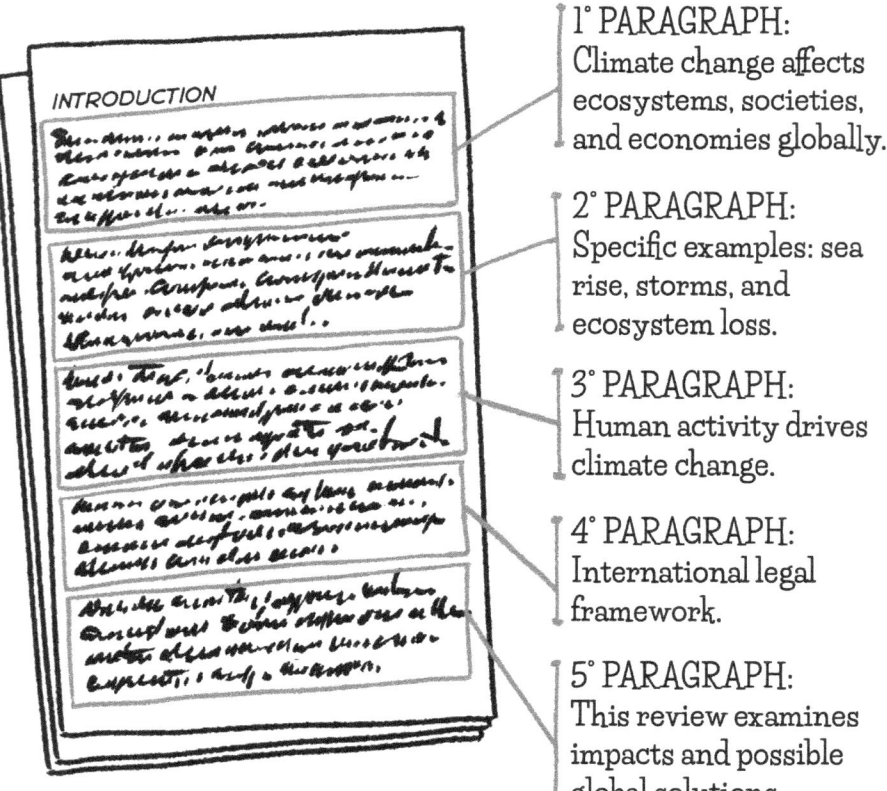

1° PARAGRAPH:
Climate change affects ecosystems, societies, and economies globally.

2° PARAGRAPH:
Specific examples: sea rise, storms, and ecosystem loss.

3° PARAGRAPH:
Human activity drives climate change.

4° PARAGRAPH:
International legal framework.

5° PARAGRAPH:
This review examines impacts and possible global solutions.

A structured approach eliminates the need to wait for inspiration. Even when you don't feel like writing, you can still make progress by filling in structured sections and refining them later. The more you rely on a systematic process, the easier writing becomes over time.

The paragraph: a blueprint for more mechanical writing

A well-structured paper is built from well-structured paragraphs. The key to writing paragraphs efficiently is to break the process into simple steps. This is the classic structure of a paragraph:

* **1st Sentence:** Topic Sentence – Introduce the main idea clearly.

* **Middle Sentences:** Evidence & Explanation – Provide data, citations, or reasoning to support the topic. It can be multiple sentences with different examples and citations.

* **Last Sentence:** Conclusion or Transition – Summarize significance (could be even a repeat of the first sentence) or connect to the next idea.

Here's a more concrete example on how to construct a solid paragraph in a systematic way:

Step 1: Outline the paragraph

Before writing, define the main idea of your paragraph (see previous section). Think of this as **the topic sentence** or the core message.

Example:

Paragraph 3: *The problem of climate change in North America.*

Step 2: Write the skeleton (Draft with references)

Start by laying out key pieces of evidence with references, even if the sentences are very rough and not connected. The goal here is to capture the logical structure.

Example:

"Climate change is currently a major problem in North America. Sentence about hurricanes (Perez et al. 2012). Sentence about increased flooding (Smith et al. 2005). Sentence about increased wildfires (Veblen et al. 2010). All this is evidence that climate change is affecting storms, flooding, and wildfires."

Step 3: Expand and refine (First draft)

Now, complete the sentences and improve readability while keeping the logical flow intact.

Example:

"Climate change is currently a major problem in North America. Since 1960, hurricanes have increased in both intensity and frequency, creating significant challenges for human livelihoods and businesses (Perez et al. 2012). Smith et al. (2005) reported that high variations in rainfall have also increased flooding events, causing economic losses in the billions of US dollars. Wildfires are another growing issue, as prolonged droughts—exacerbated by climate change—have triggered some of the largest wildfires ever recorded in North America (Veblen et al. 2010). Collectively, these changes provide strong evidence that climate change is impacting storms, flooding, and wildfires, with escalating consequences."

The **key lesson** here is that writing becomes much easier when you **first lay out the structure, then build on it.**

You can use this example as a guide for your next paper.

What is the most important section of a paper?

When I ask researchers what the most important section of a paper is, most answer **"the results section"**. This makes sense—after all, results are what make a study unique and what took us months or years to get. Each point on a graph is a test tube, a gel, or a plant we measured with effort. However, **from a peer review and publishing perspective, the results section is not the first filter—it's actually one of the last.**

THE MOST IMPORTANT PART OF A PAPER COMPETITION

The title is the first thing **everyone** sees, followed by the **abstract**. This applies not only to peer reviewers and editors but also to potential readers after publication. Keep in mind that editors and reviewers try to consider the perspective of readers when evaluating your manuscript.

The title is the first thing readers, editors, and reviewers see, followed by the abstract. This applies not only during peer review

but also after publication when attracting potential readers. Editors and reviewers assess your manuscript with the audience in mind, considering its clarity and relevance.

Consider this: **if 500 people read your title, only one of them may read your full paper**. A weak title can cause editors to reject your manuscript outright—perhaps because they assume it is too niche, lacks impact, or is out of scope for their journal. Similarly, a poorly written **abstract** may discourage reviewers or editors from engaging deeply with your work.

Where should you spend the most time?

* **Title:** much more time per word than the abstract.

* **Abstract:** Much more time per word than the rest of the manuscript.

* **Figures:** These may be the first thing editors and readers look at after the title and abstract, so they must be clear and easy to understand.

This is not only about how editors and reviewers see our paper but also about how people nowadays read papers. When writing a paper we tend to assume that people will read every single word we wrote in the order we presented them. But this is, in my opinion, very far from reality. Most people only read the title, a subgroup may then read the abstract, and see the figures (if they have access to them, title and abstract are always available). Then they may go to the first paragraph of the discussion or to the conclusion. Rarely do people read a paper from title to acknowledgements, so we need to plan for that: **if something is very important in your paper, do not bury it in the middle of the text of the paper, make that finding or idea visible to a person skimming over your text.**

Avoid jargon whenever possible

Jargon refers to technical vocabulary specific to a discipline or subdiscipline. While it can help convey details with precision, it also **alienates readers**—even within related fields or from different areas of the planet.

Studies show that **papers with jargon-heavy abstracts are read less than those written in simpler, more accessible language**. This is especially true if your research has interdisciplinary relevance. The goal is to **increase readability, not to impress readers with fancy words**.

A good compromise is to **briefly define key technical terms**.

For example:

"We studied invasional meltdown, which refers to the positive interactions among non-native species, on the mice of Isla Navarino, Chile." This allows you to **use precise terminology while still making your research accessible.**

Sadly, details matter more than you think

When we are finally ready to submit our paper, small **details in writing and formatting matter a lot**. An extra day spent refining your abstract or making your figures more readable can make a **huge** difference in how your paper is received by a journal. A typo on the abstract, a name mispelled or a wrongly cited paper can decide if your paper goes out for review or not.

There is an unspoken assumption that if you did not fix obvious mistakes that for the editor or reviewers took minutes to find, that you may have made many more profound mistakes that take longer to find. Of course, this isn't always true—many of us make typos— but it is often how evaluators interpret careless errors. Before submitting, **take your time, think carefully, and polish your work**. The extra effort is always worth it.

Many journals have very specific requirements, which can feel annoying. These requirement could make no sense to us and can include a request of a text on why your paper is a good fit for the journal in 100 words, formatting the abstract so some aspects are more clear (e.g., management implications), suggest names of multiple reviewers with their full affiliations, detailed explanation on authorship, and a never ending and growing list of things that you may not consider relevant and likely found you exhausted after days of working hard on the manuscript. You need to understand that **editors often put a lot of time into coming up with these questions, and likely they care about them.** Following these requests can make a big difference, even if they are optional.

#05

IT TAKES A VILLAGE

Many researchers think of writing as a solo activity, but the best papers often come from collaborative efforts. Getting the right co-authors involved early in the process can share the burden of writing, improve the quality of the manuscript, and even make publishing more enjoyable. Collaboration is not just about dividing tasks; it is about refining ideas, strengthening arguments, and increasing the likelihood of acceptance.

Get new collaborators. Get people to help!

Having a paper cited as "Your Last Name et al." is a great achievement, and the number of co-authors doesn't significantly diminish your merit. In fact, in most scientific fields, multi-author papers are the norm and they are more cited. It is rare today to see a single-author paper, especially in empirical research, and for good reason—science is a collective effort.

One of my biggest early mistakes was not including as co-authors

people who helped me significantly. There are many unspoken rules in academia that, if made explicit, would clearly be wrong. One of those is the idea that PhD students should limit the number of co-authors on the papers coming from their dissertation chapters—

typically including only their advisor and people with very specific, irreplaceable roles. This belief led me to exclude two key people who truly deserved authorship in my PhD work.

The first was Romina Dimarco (a researcher and my wife), who played a fundamental role in my experiments. She helped me polish ideas, collect data, structure my manuscripts, and provided emotional support throughout the process. Yet, despite all of this, she was not listed as an author on any of my PhD chapters. The second was Nate Sanders, one of my greatest mentors and a professor at the University of Tennessee back then. He helped develop my ideas, refine experimental designs, attended all my talks, and read in great detail every one of my manuscripts. More importantly he was incredibly supportive. But for reasons I still can't understand, I didn't include him as a co-author either.

Both Romina and Nate did exactly what I asked them to do, they seemed happy to help, never asked to be authors, and were really happy to know when those papers got published. But looking back, it would have been more fair and more logical to recognize their contributions formally. Science is a collective process, and authorship should reflect that.

A note of caution: Once you tell someone they will be an author, you cannot take it back. There is no way to "unsay" it without creating conflict. To avoid misunderstandings, it is wise to set expectations early on. Let potential co-authors know what kind of contributions are expected and what level of involvement they will need to have. At the same time, it is important to be generous—minor contributions should not necessarily disqualify someone from authorship. In most cases, I'd rather have a co-author who contributed minimally than start an unpleasant conflict over authorship. In my experience, these battles are often more damaging than simply adding a name to the author list. Anyway, be sure that the people that enter the author list deserve it, they are in the correct place in the author's other, and that no one that deserves to be an author is excluded. Also be sure that all authors approve the paper and understand the responsibilities that being an author imply.

Choosing the right collaborators

Not all collaborations are equally productive. Technically, the best collaborations happen when people bring complementary skills to the project. Writing a scientific paper requires multiple skills—conceptualizing research questions, designing experiments, analyzing data, interpreting results, making figures, structuring a manuscript, and refining the language. Few people are equally strong in all of these areas, so choosing co-authors who can fill in the gaps can be incredibly valuable.

If you are a strong writer, it can be useful to collaborate with someone who is excellent at data analyses or visualization. If you struggle with structuring your manuscript, working with someone who has an intuitive sense of logical flow can be a game-changer. One of the simplest ways to assess your own strengths and weaknesses as a writer is to ask yourself: Which part of the writing process do I find easiest? Which part do I avoid? The answers can guide you in finding collaborators who complement your skills.

Collaboration with people with your same skills can be also very important. Having someone that is there when you need them to talk about some details and that is happy to pick up with the writing when you are exhausted, can be super important. Something to keep in mind is that collaboration should be mutually beneficial. Nobody wants to feel like they are doing all the work while others get credit. The best scientific teams develop a natural rhythm of contribution, where each member adds value in their own way.

How collaboration improves writing

One of the greatest advantages of collaboration is that fresh eyes can catch weaknesses that you may have overlooked. When one is working on a manuscript for weeks or months, it's easy to become blind to logical gaps, unclear sentences, or missing references. A good collaborator will spot these issues quickly, saving you from submitting a manuscript that may not be as strong as you think.

Collaboration also makes writing faster and more efficient. Writing an entire paper alone can be slow and exhausting. When multiple people are involved, different sections can be drafted simultaneously. One person might focus on the introduction, while another refines the methods, and yet another improves the figures. This division of labor makes the process more manageable and less overwhelming.

Beyond efficiency, collaboration also improves the intellectual quality of the paper. Science thrives on diverse perspectives, and a well-rounded research team can offer different interpretations of the data, refine arguments, and ensure that conclusions are robust. A biologist, a statistician, and a geographer might look at the same dataset in different ways, and integrating those perspectives can lead to a much stronger and interesting final product.

Another benefit of collaboration is that it can increase the likelihood of getting published. Papers that have gone through multiple rounds of internal feedback—from co-authors, mentors, and colleagues—are often much stronger when they reach peer review. Reviewers are less likely to reject a paper that has already been refined by multiple critical eyes.

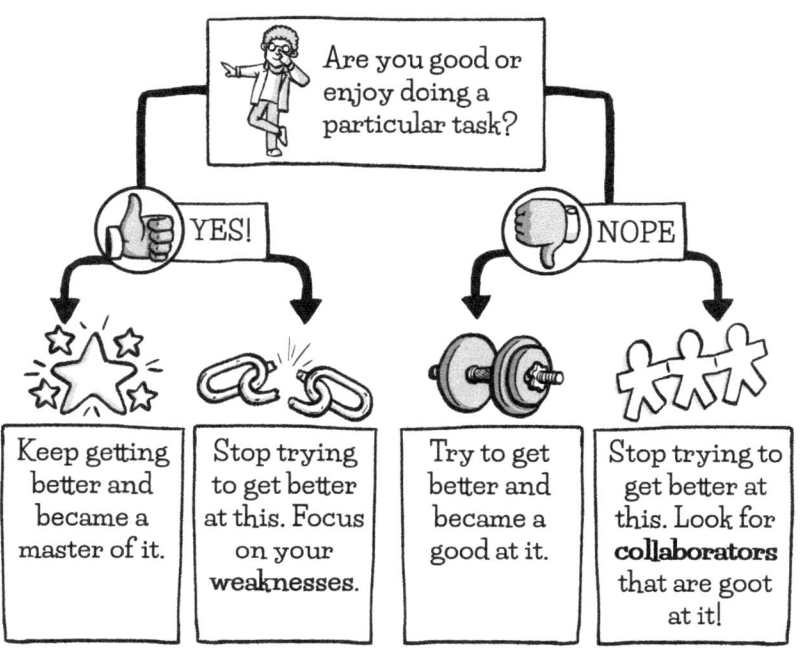

#06

STORYTELLING IN SCIENCE

Scientific writing is often seen as dry, rigid, and purely factual, but at its core, a **good scientific paper is a story**. It has a beginning, a middle, and an end. It has a central message that holds everything together. And like any well-told story, it must be clear, logical, and engaging.

A well-structured paper guides the reader from the introduction to the conclusion in a seamless progression. It is not just a collection of results but a **cohesive narrative that makes sense from the title to the last paragraph**. This chapter explores how to frame research as a story, ensuring clarity, coherence, and impact.

The paper as a story

Every scientific paper should be a cohesive piece of writing. The different sections—introduction, methods, results, discussion—should work together, not as disconnected parts but as elements of a single narrative. Anything that does not contribute to this story should be **revised, reduced, or removed**. I think that there is something evolutionary about how humans process information. Stories are much more effective than random pieces of data. Having in mind what the story is that you want to tell is fundamental for a paper.

Many things can disrupt this cohesion. One common problem is including **data that are unnecessary for testing the hypothesis—** perhaps because they were collected but did not turn out to be relevant. Another issue arises when different authors write different sections, leading to inconsistencies in style and argument flow. These challenges can be overcome, but they require careful editing and, often, **a shift in mindset about how we approach writing.**

Please do not think that by a "story" I imply writing fiction or removing parts that do not fit my hypothesis. I do not mean that at all. Some research gets complex results and our paper must say that explicitly. What I refer to here as a paper without a story, is a paper that you read and don't know what the take home message is, a paper that we cannot summarize for our colleagues. A mere collection of results. Papers with data that is not relevant to test my hypothesis. We need clear stories to get our papers published and to get them to have an impact.

One simple way to **test** whether a paper has a clear story is to check for **unnecessary information**. Ask yourself: *Does this section, paragraph, or figure add to the main narrative of the paper?* If the answer is no, then it likely needs to be reworked or removed. **I do this by writing first the "abstract".**

Abstract as the starting point

One of the best ways to ensure a is to **write the draft of the abstract first**. This may sound counterintuitive—many people think the abstract should come last, summarizing a finished manuscript. But drafting the abstract early helps define the **scope and focus** of the paper before getting lost in details.

When I start writing a paper, I draft a quick abstract to get a sense of what the key message should be. This immediately helps clarify **what belongs in the paper and what doesn't**. If I find myself writing two full paragraphs in the introduction about a topic that is not even mentioned in the abstract, that's a sign that I may not need those paragraphs. If I have an entire section of results with two figures on something that doesn't appear in the abstract, I reconsider whether those results are central to the story.

At the same time, the abstract can also reveal **what's missing**. If I mention an important conclusion in the abstract but don't have a strong discussion about it in the paper, that's a sign that I need to expand on that point. Writing the abstract first is like setting a roadmap—it ensures that all sections align with the core message of the paper.

Unlearning what you were taught about writing

Many of us were taught writing rules in school that are not only **useless in scientific writing** but can actively make our papers worse. The rigid five-paragraph essay, the idea that long words and complex sentences make writing sound "smarter," or the focus on **filling pages rather than making points**—all of these habits can get in the way of clear, effective scientific communication.

In my case, I had a unique advantage: **I never wrote much until I had to write scientifically.** When I defended my Master's thesis, my mother (who had just met my advisor for the first time) asked him, **"How did you make Martín write something?"** While this was her way of making a joke at my expense, and not in the ideal moment, it was also somewhat true—I didn't write much before my Master's, and certainly not before my late years in the program.

At first, my lack of training in writing seemed like a problem. But in hindsight, **I think it was an advantage.** Because I never took formal creative writing courses or any writing classes, I never had to **unlearn** anything. My writing education came entirely from **scientific writing**, and I was never burdened with unnecessary rules that didn't apply to scientific papers.

For many PhD students and early-career researchers, the biggest challenge is not **learning** how to write scientifically but **unlearning** the habits they were rewarded for in school. Overly polished, vague introductions, unnecessary complexity, wikipedia level writing, and filler phrases need to be stripped away. For many, the best scientific writing is **short, direct, precise, and structured like a logical story**.

#07

QUICK TIPS FOR THE DIFFERENT SECTIONS OF A PAPER

Scientific papers have a **standard structure** for a reason—they guide the reader through the research in a clear and logical way. However, writing each section effectively requires more than just following the format. Small details can make a significant difference in how well your paper communicates its message. Understanding the purpose of each section and how to write it well will improve the clarity, readability, and impact of your work.

A clear example in my field is how many early-career researchers start their title, abstract, or introduction by focusing on a **particular species or a very niche topic**. While this might be scientifically valid, **editors and reviewers may not like it**. They prefer broad statements that connect to larger, well-known issues. For example, you'll often see papers begin with:

"Climate change is one of the most pressing challenges faced by humankind."

Is this an original statement? No. Does it make sense scientifically? Maybe. But more importantly, does it follow the expected style? Yes.

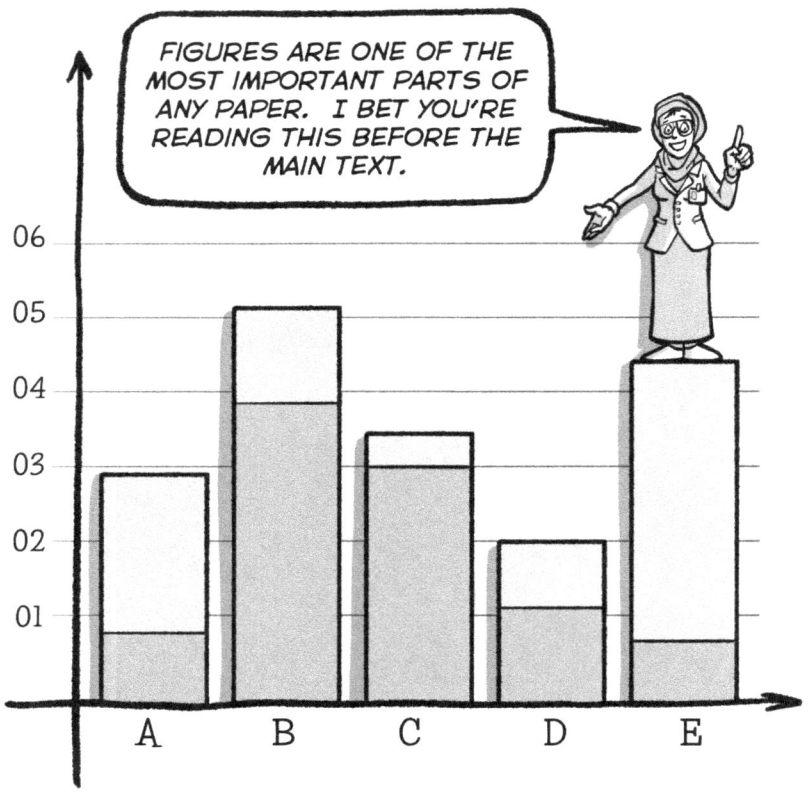

There are many things in publishing that **don't seem logical or fair, but they are part of the game.** If you want your paper to be accepted, **study the style of papers that get published and follow that model.**

Title: the first and most important impression

As seen in chapter 4, the title is arguably the most important part of a paper because it is the only thing most people will read from your work. A well-crafted title should immediately tell the reader the main takeaway of your study. I personally prefer titles that clearly state the main finding, such as *"Water is needed for plants to grow: A review of current evidence."* Some researchers prefer question-based titles, such as *"Is water important for plant growth? A review of current evidence."* Others take a descriptive approach and simply state what was done, such as *"A systematic analysis of the role of water (H2O) in plant growth."* And others may think of a funny/humoristic title.

While all styles are commonly used, I prefer the first style with the main message, because it guides the reader immediately to the study's conclusion. While getting people to read your full paper is ideal, if they can grasp the key message from the title alone, that is still a win. Question-based titles work well when results are uncertain or context-dependent, while purely descriptive titles may not attract as many readers since they do not emphasize the main finding.

Abstract: the most important paragraph. Period

The abstract is one of the most critical sections because it is often the only part people read beyond the title. The purpose of an abstract is to quickly summarize the key findings of your study. It should not waste space with generic statements like *"In this paper, we explain the role of water in plant growth. We reviewed 2,000 papers and found conclusive results."* Instead, it should directly state what was found.

The faster a person understands the main ideas of a paper, the better. If a researcher satisfies their curiosity by reading just the abstract, that is a success, a win. After all, no one gets paid based on how many words of their work are read. The goal is to communicate knowledge effectively, not to force people through unnecessary text.

Writing abstract in multiple languages also has many benefits. It mainly becomes easily accessible to many more people, and even for the people that may need it the most sometimes. Also, it helps researchers find your work. A student in Brazil may look for keywords in portuguese.

Keywords: a small detail that still matters

Keywords originated when papers were cataloged manually, but they still serve a purpose today. Many journals require them, and they help improve the visibility of your paper in search engines and databases. The key rule for choosing keywords is to avoid repeating words already present in the title or abstract. Instead, select terms that are relevant to your study but not explicitly mentioned in the main text. This can include technical jargon, synonyms, or broader concepts related to your work.

Introduction: setting the stage for your research

The introduction provides background and frames the research question. While the length may vary, a well-structured introduction follows a clear pattern. The first paragraph should introduce the broad topic and its significance, placing the research into a wider context. This is the big-picture section that explains why the general topic is important, even though your study is only addressing a small part of it.

After establishing this broad context, the introduction should gradually narrow its focus. This is where previous research is introduced, along with any knowledge gaps, conflicting findings, or unresolved issues in the literature. The goal is to create a natural lead-in to the specific research question being addressed.

The final paragraph of the introduction should clearly state the study's objective. It should explain exactly what the research aims to do, often in the form of specific research questions or hypotheses. A strong introduction does not simply list background information but guides the reader toward understanding why the study was necessary and what it set out to accomplish.

Methods: can someone redo your study?

The methods section should be the easiest to write because it is a straightforward description of what was done. The key goal is reproducibility—after reading this section, someone else should be able to replicate the study. If the methods follow a standard protocol, there is no need to describe every detail. Instead, previous studies can be cited to avoid redundancy, as in *"We followed the protocol proposed by Hui et al. (2003) to sterilize our soils."*

Methods sections can often be dry, which can make it difficult for readers to fully grasp what was done. One way to make this section clearer and more engaging is to include figures, diagrams, or photographs of experimental setups. Humans process visual information quickly, so adding a clear illustration can make the methods much easier to understand.

Results: the data

The results section consists of two key components. First, there is the text that describes the findings in detail, typically accompanied by statistical analyses. This part is often technical and can be challenging to read, but organizing it well and using subheadings can make it easier to follow.

The second, and often more impactful, part of the results section is the figures. Many readers will skip the text entirely and go straight to the graphs. This is why it is crucial to ensure that figures are clear, well-designed, and visually appealing. A well-made figure can convey results more effectively than a paragraph of text.

When designing figures, it is useful to think about whether they could be used in a university lecture. If a graph is simple, clear, and visually intuitive, it will be easier for both scientists and students to understand. Another helpful trick is to make figure captions more descriptive. Instead of simply stating "Figure 1: Relationship between water addition and plant growth", it is better to write "Figure 1: Plants with more water grew significantly taller." A small change like this makes it easier for readers to interpret the findings at a glance.

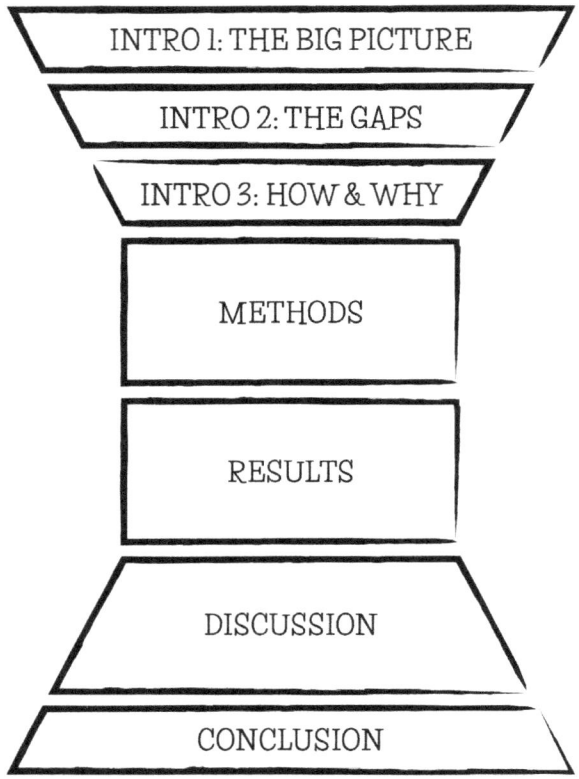

If a result is important, it should be in a figure. If that is not possible, it should be in a table. Sometimes small tables that are well designed can be equally as good as figures to share effectively a message. Avoid burying key findings deep in the text where they may be overlooked.

Discussion: explaining your results and the bigger picture

The discussion section is where the results are interpreted and placed into context. Unlike the introduction, which sets up the research question, the discussion explains what the findings mean. A common mistake is starting the discussion with a general summary of the research problem. Instead of writing something vague like *"Water has been proposed as a major factor in plant growth, but a comprehensive review has been lacking,"* it is better to begin with the main result: *"In agreement with previous research, we found that water enhances plant growth in most circumstances."*

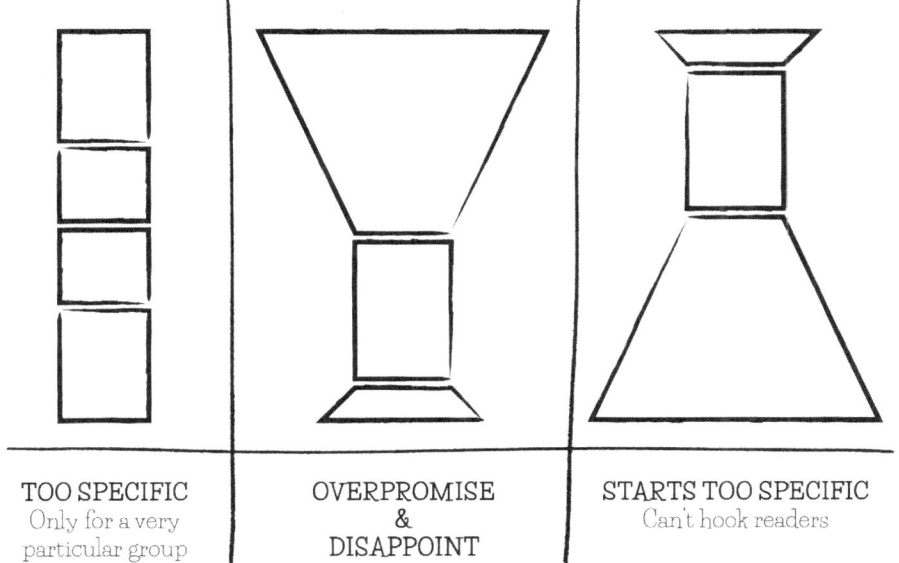

TOO SPECIFIC
Only for a very
particular group

OVERPROMISE
&
DISAPPOINT

STARTS TOO SPECIFIC
Can't hook readers

A well-structured discussion follows a logical order. It should start with the study's main finding, then compare it with previous research, explore possible explanations, acknowledge limitations, and suggest directions for future work. Keeping the discussion organized in the same order as the introduction, methods, and results makes the paper easier to follow. Readers should not have to jump back and forth between sections to understand how everything fits together.

Clearly articulating the study's limitations while demonstrating that they do not undermine the validity of your conclusions is an effective way to preempt overly critical reviewer comments (from reviewer #2 typically:)).

There are plenty of places to be creative in scientific writing. From solid lines on the importance of your work. To make aesthetically beautiful graphs to have a great title. But you don't want to depend on your creativity to get your paper down. We need to find uncreative mechanical tools that help us get there. If you get creative, wonderful! But if you don't you will publish anyway. Build your own template of how a paper should look like

Writing the first draft of a scientific paper: a generic model

I still think that copying the structure of a paper that is similar to your and that you liked a lot, is better. But here is a generic model that can be of use to come up with a first draft.

Introduction

Establishing the research context

*   The Bigger Picture – Scientific research contributes to solving broader societal and environmental challenges by expanding knowledge and informing policy. Positioning your study within this context highlights its significance beyond the immediate research question.

*   Specific Research Focus – Narrow the scope by explaining how the topic of your study addresses a specific aspect of the broader issue that is needed.

Literature background and research gap

*   Existing Knowledge – Summarize how prior studies have approached your research topic.

*   Identified Weaknesses/gaps – Highlight knowledge gaps, inconsistencies, or limitations in previous research.

*   Synthesis of Weaknesses and/or gaps – Integrate these shortcomings to define a clear research gap that your study aims to fill.

Research question and objectives

*   Defining the Research Gap – Explain how the weaknesses in prior literature lead to your research question.

*   Stating the Research Question/hypothesis – Clearly articulate the main question or hypothesis that your study addresses.

*   Research Objectives – Outline the specific aims and hypotheses tested in your study.

Methods

Overview of study design

* Approach – Describe how your methodology generates data relevant to answering your research question.

* Key Study Components – Summarize essential elements (e.g., data collection, sampling strategy, experimental setup).

* Addressing Limitations (if needed)– Acknowledge methodological constraints while justifying the reliability of your approach.

Detailed methodology (think in someone that needs to re do your experiment)

* Describe each procedural step in a logical sequence.

* Justification of Techniques (if needed) – Explain why specific methods were chosen over alternatives.

* Repeat for each procedure.

Results

Structuring the Results

Provide a concise, non-interpretative overview of each of the main results following ideally the order presented in the methods.

* Key Data Representation – Introduce figures, tables, or statistics that support your findings.

* Interpretation of Visual Data – Highlight trends, patterns, or anomalies shown in figures and tables. Make figures as clear and nice as you can.

* Repeat for each additional result.

Discussion

Interpreting the results

* Present the main results (e.g., In this study we show that...). Connecting Findings to the Research Question – Explain how your

results contribute to answering the central research question.

* Comparison with Existing Literature – Discuss how your findings align or contrast with prior research. Follow the structure from the result section if possible.

* Addressing Study Limitations – Acknowledge the constraints of your study and their potential impact on interpretation (very important to prevent reviewers comments).

Broader implications

* Generalization of Findings – Explain how your study contributes to broader scientific understanding.

* Practical or Theoretical Impact – Discuss how your findings

* can be applied or how they advance scientific theory.

* Future Research Directions (optional) – Identify open questions and suggest areas for further study.

Conclusion (optional)

Summarizing the study

* Restating the Research Purpose – Briefly summarize the research question and objectives.

* Recap of Key Findings – Highlight the most significant results in relation to the research question.

Final takeaways

* Filling the Research Gap – Explain how your study contributes to advancing knowledge.

* Connecting to the Motivation – Relate the conclusions back to the broader problem outlined in the introduction.

What if you still can't write? Lead the writing!

Even with all the mechanical writing strategies outlined in this book, you may still struggle to produce a beautifully written page. And that's perfectly fine—scientific writing is not about literary elegance; it is about clarity and structure. By following these steps and working with collaborators, you should be able to produce something that, with revisions and feedback, can reach the stage of publication.

However, if you truly find yourself unable to write—whether due to writer's block, lack of time, or simply not enjoying the process—there are solutions. One of the most effective is **collaborative writing**. In science, there is no need for ghostwriters like in the book industry; instead, you can bring in co-authors to help with the process. Many successful papers are written through a division of labor: one person takes on the introduction, another tackles the discussion, and someone else polishes the methods and results. If writing is your obstacle, **become the organizer**. Ask a colleague, postdoc, or student to take the first pass at a section. Once something is on the page, it becomes far easier to refine and complete.

Certain papers can also be emotionally or psychologically difficult to write. This often happens when a paper carries personal significance — perhaps it represents **years of hard work, a major career milestone, or a finding you hope will earn you recognition**. The weight of these expectations can make the writing process overwhelming. If that happens, acknowledge the difficulty but find ways to **get the paper written anyway**. Use all the tricks available —collaboration, outlining, or setting artificial deadlines—but do not let the project stall indefinitely.

I have personally faced this situation. There was one paper that lingered on my to-do list for **ten years**. The data were solid, but it never excited me enough to prioritize writing it. Eventually, I had a student who needed writing experience and had time to work on papers. That was the perfect opportunity—I handed over the project, this student led the writing and the paper finally got written. Everyone benefited: the student gained experience and a publication, and I finally got the paper off my list.

If you cannot write a paper yourself, find someone who can help. Whether through collaboration, delegation, or simply forcing yourself through the process, the goal is to **get the paper written**. A paper that exists—even in a rough draft—is infinitely better than a perfect idea that never gets published.

Scientific writing does not have to be a lonely process. Collaboration not only makes the process easier, but it also leads to better papers and, ultimately, better science. The key is to involve the right people early in the process—not just at the end when the manuscript is nearly finished. Being strategic about co-authorship is important, but being generous with credit is just as important. Science is built on relationships, and giving due recognition to those who help you along the way strengthens those relationships. A well-chosen team of collaborators can make writing faster, less stressful, more fun and more intellectually rewarding.

What to do if your english is really bad?

I am happy to discuss topics where I am a true expert! My first recommendation is to **study English**. Becoming proficient in a language is always valuable—not only will it improve your reading and writing skills, but it will also help you appreciate literature, music, and other cultural aspects. There are many effective online tools, such as **Duolingo**, that can help.

As I mentioned before, **collaboration** can also play a key role in improving your English. Adding a **native speaker as a co-author** is a common solution. However, in my view, **language editing alone is not a sufficient reason to grant co-authorship**. If the person has contributed to the ideas or other sections of the paper, then authorship is fully justified.

Another option is to **find a good Samaritan** willing to help you polishing your text. If you live in a country where English is not the primary language, this might be more challenging. However, I have successfully helped researchers connect with people willing to assist with English via **social media**. You can post something like: *"I am a researcher from* [Your Country]*, and I am struggling with the English grammar of my paper. Can anyone help?"* Tag me, and I will try to assist! For real, tag me and I will help!

In my opinion, **good English is much less important than good writing structure**. A clear, logical outline and well-structured paragraphs make up **90% of a strong paper**. Fixing minor or major language issues is something a fellow scientist—or even AI—can handle in just a few hours. That said, **improving your English and seeking help when needed are both great strategies before you submit the paper.** Very sadly there are conscious or unconscious biases against people with poor english. So having a nearly perfect English in your text can reduce this problem.

On getting help from writers who are not scientists...

Many people struggling with writing look for resources in good writing from other areas. You may have a good friend that writes for newspapers, have in your family a professional writer, a certified translator or a writing teacher. All this can be a good resource if used properly, but also sometimes can be a true disaster.

In Latin America many people write their papers in Spanish and then get a paid translator to write the version to be submitted. It is amazing to see these papers and how the wrong use of some fundamental words can make the work look like it was written by someone that is completely clueless. There are many idiosyncrasies about academic writing (e.g., the shorter you can write the better!). A typical problem among non-academic writers is their habit of avoiding repeated words in a paragraph, as if they were allergic to repetition. Doing that clearly makes the text harder to read, but in science many times we do not have proper synonyms for a word. Using words that may seem as synonyms interchangeable can create a nightmare for the reader. In my research the exact area where I do my field experiments can be called "plot", "site" or "parcel", but I cannot change the term, otherwise it becomes impossible to understand. Remember that clarity is a #1 goal of scientific writing.

As an editor I sometimes see these problems where some fundamental words are not used properly and that makes me doubt the competence of the authors…

Using reference management tools

There are many **reference management software programs** that help you organize and cite references efficiently. These tools are incredibly useful for researchers, offering several advantages. They allow you to systematically **store, manage, and retrieve references,** ensuring that every source you read is easily accessible for citation later. They also give your paper a **professional and polished look in seconds**, as references are automatically formatted according to journal requirements (though occasional corrections may be needed).

I personally **love reference management tools**, especially when I need to **resubmit a paper to a different journal after rejection—** remember the hashtag #rejectionistherule. With just a few clicks, you can reformat your references to match a new journal's style, saving time for a really rewarding effort (formatting can be painful!).

These tools also offer other benefits, such as **collaborating on shared libraries, storing PDFs, and creating "to-read" lists** to keep track of relevant literature.

There are three major reference management programs that function in similar ways. When I started, **EndNote** was the only widely used option, and while it remains a great tool, its major drawback is that it **is not free**. Fortunately, **Zotero** and **Mendeley** are free alternatives that now work just as well. Each software has its own strengths and problems, so if you enjoy exploring different tools, try them out and see which one suits you best. If not, simply ask your colleagues what they use and follow their lead. **Using the same software as your collaborators makes it much easier to share references and get help with troubleshooting.**

First things first: The importance of a proper literature review

In this book I do not focus on how to do your research, but as Newton famously said, "If I have seen further, it is by standing on the shoulders of giants." Science progresses by building on previous work, and to contribute meaningfully, we must know—and properly cite—the foundational research in our field. A thorough literature review is the key to achieving this and **it needs to be clear in your paper that you know the literature**. Our goal should be to understand the most influential work in our discipline, but given the sheer volume of papers published every month, this is no easy task. Fortunately, there are several tools to help us.

A good starting point is simply asking experienced researchers—such as your advisors or PI—which papers they consider essential reading. These "must-read" papers provide a foundation, and from there, you can explore who has cited them and what references they include, helping you map the intellectual landscape of your topic.

Beyond personal recommendations, search engines and bibliographic databases are crucial for a comprehensive literature review. The gold standards are **Scopus** and **Web of Science**, which index papers from well-established journals and offer precise, repeatable searches. When you master the use of the proper keywords, it gets really fun. However, they have limitations: they are not free, and they exclude a significant portion of scholarly work, such as many books, book chapters, non-English papers, and other resources that may be highly relevant depending on your field.

Another valuable tool is **Google Scholar**, which operates quite differently. Unlike Scopus and Web of Science, it is free and covers nearly everything ever posted online. However, its biggest drawback is that it lacks systematic search functions, making it difficult to conduct precise, repeatable searches. Despite this limitation, it remains a powerful way to discover important and sometimes overlooked research.

Failing to conduct a proper literature review can seriously undermine your credibility. If you overlook key studies, others—especially reviewers—may assume you lack knowledge of the field. Please note that reading the title and abstract is not enough to be able to cite the paper, this can lead to embarrassing citation errors. Additionally, knowing the major authors in your area is crucial since they may very well be the ones reviewing your paper!

By taking the time to conduct a thorough literature review that is reflected on the citations in your manuscript, you ensure that your work is well-grounded, relevant, and positioned effectively within the ongoing scientific conversation.

VERY QUICK CHECKLIST FOR BEFORE SUBMITTING A SCIENTIFIC PAPER

Title

☐ Does the title clearly convey the main finding of your study?

☐ Is the title engaging, informative, and aligned with the journal's style?

☐ Have you avoided unnecessary jargon or overly technical phrasing?

Abstract/keywords

☐ Does the abstract summarize the key results in a clear and concise way?

☐ Have you avoided filler phrases like "This study investigates…" and instead jumped straight to the findings?

☐ If applicable, have you included an abstract in another language to improve accessibility?

☐ Are your keywords distinct from words already present in the title and abstract?

Introduction

☐ Does the first sentence introduce a broad and widely recognized issue rather than a niche topic?

☐ Have you provided a logical progression from broad context to specific research gaps?

☐ Is it clear that you know (and cite) the key literature on the topic?

☐ Does the introduction clearly state the question(s) or hypothesis being tested?

☐ Have you justified why your study is necessary without simply listing background information?

Methods

☐ Is the methods section detailed enough for another researcher to replicate your study?

☐ Have you cited standardized protocols instead of unnecessarily explaining well-known methods?

☐ If applicable, have you included diagrams, photos, or visual aids to clarify complex methods?

☐ Have you kept the methods concise by focusing only on what is essential to reproduce the work?

Results

☐ Are the results presented in a logical order that follows the study's main objectives?

☐ Have you used figures and tables to visually represent key findings rather than relying solely on text?

☐ Do all figures have clear captions that summarize the main takeaway rather than just describing the content?

Figures and Tables

☐ Are the titles and captions detailed enough for readers to understand the data without referring to the main text?

☐ Have you kept figures as simple as possible while still conveying all necessary information?

☐ Have you ensured that all visuals meet the formatting guidelines of the target journal?

Discussion

☐ Do you begin by stating the key findings and explaining their significance?

☐ Have you compared your results with existing research?

☐ Does the discussion follow the same logical flow as the introduction and results sections?

☐ Have you acknowledged limitations while demonstrating that they do not invalidate your main conclusions?

☐ Are future research directions clearly outlined without making vague or unrealistic claims?

General writing and Submission

☐ Have you reviewed the author guidelines for the target journal to ensure formatting compliance (e.g., word number, citation style, special requirements)?

☐ Did you write a solid letter to the editor?

☐ Did you take your time to complete all the questions that are asked in the portal for submission?

#08

THE PEER REVIEW PROCESS

The peer review process is one of the defining characteristics of scientific publishing. It serves as a **filter** to ensure that only well- supported, rigorous, and original research makes it into the literature. However, peer review is not just about gatekeeping—it is an opportunity to **improve your work** by incorporating expert feedback. Understanding how reviewers evaluate papers and how to respond to their comments effectively can make the process smoother and increase the likelihood of publication.

Three different layers of evaluation in peer review

Peer review consists of three distinct layers of evaluation, each serving a different purpose in assessing a manuscript's quality, relevance, and suitability for publication. Understanding how each stage works can help authors navigate the process more effectively.

The first layer is the **Editor-in-Chief review**, where the journal's head determines whether the manuscript aligns with the journal's scope and standards. Many papers are rejected at this stage due to a poor fit with the journal's focus, weak writing, or a lack of novelty. Since this is often a **quick initial assessment**, it is crucial for authors

to ensure that their paper is well-structured, clearly written, and framed in a way that highlights its significance. Something to keep in mind is that the first filter can be done very quickly (~10 minutes), and by someone that may not be an expert on your field, so avoiding jargon and being extra clear on the implications are fundamental for this stage.

The second layer is the **Associate Editor assessment**. If the Editor- in-Chief deems a paper suitable for further evaluation, it is assigned to an Associate Editor, typically an expert in the field. The Associate Editor evaluates the manuscript's **scientific validity, clarity, and overall contribution**. Their primary role is to check for its novelty and quality in a way the Editor-in-Chief (EIC) cannot do. Then if everything looks good, the associate editor **identifies appropriate peer reviewers** and oversees the review process, though they may also desk-reject the paper (remember that the EIC is unlikely an expert on the topic).

The final filter is the **peer reviewers' evaluation**, where independent experts assess the study's validity, clarity, methodology, and contribution to the field. Reviewers provide feedback that can range from minor suggestions for improvement to major concerns requiring extensive revisions. Their assessments inform the Associate Editor, who is the person that typically makes the call if a paper needs to be rejected, revised or accepted. Then the Editor-in-Chief checks for possible issues in the evaluation and makes the final decision on whether the paper should be accepted, revised, or rejected.

Each stage of the peer review process plays a very different and crucial role in ensuring the quality of scientific publishing. Recognizing the expectations at each level can help authors **better prepare their submissions, respond to feedback effectively, and improve their chances of publication.**

LEVEL 1: EDITOR IN CHIEF

LEVEL 2: ASSOCIATE EDITOR

LEVEL 3: 3 REVIEWERS

How to respond to the editors and the reviewer comments

Every researcher will, at some point, receive **critical comments from reviewers**. Some will be constructive and insightful, while others may feel frustrating or even unfair. Regardless of tone, every comment should be **addressed professionally and systematically**. Also, do not forget to respond to the comments made by the Associate/handling editor. The Associate/handling editor is the person who leads the review process, expends substantial unpaid time on it, and who typically has the bigger role in deciding if a paper is accepted or rejected. Very rarely an editor in chief will overrule a decision made by an associate editor.

The best strategy is to **categorize** reviewer comments and prioritize changes based on their **difficulty and necessity**.

1. **Easy to fix and necessary:** These are the straightforward revisions—clarifying a sentence, adding a missing citation, or correcting an error. There is no reason to resist these changes.

2. **Hard to fix but necessary:** These require more effort, such as reanalyzing data, adding new discussions, or restructuring sections. While they may be challenging, addressing them strengthens the paper.

3. **Easy to fix but unnecessary:** Some suggestions may not be crucial but can be accommodated without harming the paper. Making these minor changes can help avoid unnecessary friction with reviewers.

4. **Hard to fix and unnecessary:** If a reviewer suggests an analysis or change that is irrelevant or weakens the paper, this should be politely explained and justified rather than ignored.

My personal rule is: **If I can make the change (even if I know that is not so important) I will do it. This does not only apply to reviewers from a journal, but also to the people that I ask to make comments on my manuscript.** Incorporating reviewer feedback is crucial not just for publication but as a way to acknowledge the time and effort that reviewers invest in evaluating the work. Many times, their fresh eyes reflect much better what a random reader will understand, so following the comments can add a lot. A well-crafted response that demonstrates engagement with their suggestions can influence the editor's final decision. On the other hand, a poorly written reply letter, where you ignore comments, attack reviewers, or don't give enough detail is a clear path to rejection.

If you get a chance to send a new version, this is always good news. The editor thought that if you are able to fix the issues highlighted, then you are published. If you reach this stage, the decision is mostly in your hands.

When you don't need to make reviewers happy

Not all reviewer comments are helpful or appropriate. Some may be **unreasonably critical, dismissive, or even biased**. There are three key cases where it is acceptable to **push back against reviewer comments**:

1. **When the reviewer is factually wrong.** If a reviewer misinterprets the study or asks for changes that contradict established methods, it is reasonable to clarify why their suggestion is incorrect. Always assume good faith, and assume that it is your fault for not being extra clear. Clarify what needs to be clarified, but do not follow the reviewers suggestion.

2. **When the comment would make the paper worse.** Some suggestions may not be intrinsically wrong, but can introduce unnecessary complexity or distract from the main findings. In such cases, responding with a clear and well-reasoned explanation can be more beneficial than simply making the change.

3. **When the comment is rude, inappropriate, or discriminatory.** Unfortunately, some reviewers use unnecessarily harsh or dismissive language. If a comment is offensive or biased, it should be reported to the editor rather than engaged directly.

Even when a suggested change seems unnecessary, **it can sometimes be useful to make the revision anyway**—especially if it helps demonstrate that the original approach was stronger. For example, if a reviewer asks for an additional analysis that is not essential, conducting it and showing that the conclusions remain unchanged can reinforce the robustness of the original findings (and you may choose not to even report in the paper this new analisis or add it to the appendix).

It is also useful to remember that reviewers represent the general opinion in your field. If multiple reviewers highlight the same issue, it is likely that other readers will have the same concerns.

Great papers can get rejected

Rejections happen for many reasons, and not all of them have to do with the quality of the paper. One of the most common reasons is that the paper **doesn't fit the journal's scope**. This may seem like a straightforward issue, but journal scopes are often shifting. Editors may decide that they are publishing too much on a certain topic and want to make room for other subjects, or they may want to establish leadership in a new field.

Another reason for rejection can be **timing**. If too many papers on a similar topic have recently been published, the journal may reject a submission simply to avoid redundancy. Even logistical issues, such as difficulty finding qualified reviewers, can contribute to rejection. Some editors justify this by arguing that if no one is willing to review a paper, there may not be enough interest in the community to justify publishing it—though this logic is flawed, particularly when one is considering papers from underrepresented regions or lesser-known authors.

Mood and bias also play a role in rejections, even though this is rarely discussed. If an editor or reviewer happens to be in a bad mood when evaluating a paper, they may be harsher than they otherwise would be. Science is often portrayed as a rigorous and objective process, but the reality is that it **is conducted by humans, and humans are not perfectly rational.**

For all these reasons, rejection should not be taken personally. Having multiple papers rejected does not make a researcher look bad—what matters is persistence and moving forward.

The letter to the editor: an overlooked opportunity

When submitting a manuscript, many journals allow or require **a cover letter to the editor.** These letters are sometimes ignored, but they can still serve as **a final opportunity to convince an editor to send the paper for review.**

A common mistake in cover letters is **repeating the abstract** in different words. Editors will already read the abstract, so simply rephrasing it adds no value (and may even upset a busy editor). Instead, the letter should highlight **why the paper is a strong fit for the journal and why the editor should consider it seriously.**

Strong cover letters typically mention three key points:

* why the journal's audience is ideal for the paper,
* how the study contributes new knowledge, and
* why the journal is well-positioned to handle it (e.g., great editorial board.

If possible, referring to relevant editorial statements or previously published papers in the journal can further strengthen the case, and shows that you are familiar with the journal and part of its community, which is a big plus.

A particularly damaging mistake but very common (!) is **getting the journal name wrong.** If the letter is copied from a previous submission and still contains the name of the rejected journal, it signals to the editor that the paper was already rejected elsewhere and that the author did not take the time to update the submission properly.

How to suggest reviewers (and whom to exclude)

When submitting a manuscript, many journals allow authors to suggest potential reviewers. While this is optional, providing reviewer suggestions can help ensure that the paper is evaluated by experts who **understand and appreciate its context.**

A common mistake is n**ominating only the biggest names in the field.** These researchers receive an overwhelming number of review requests and often decline. Unless you have interacted with them personally, you cannot reliably predict how they will review your paper.

Another misconception is that it is beneficial to **suggest close friends as reviewers**. Most will decline due to a conflict of interest, and those who accept may provide **weak or overly positive reviews that editors will disregard**. A good Associate Editor will often check whether you have collaborated with the suggested reviewer, or may infer a close connection from the tone of the review. Attempting to game the review process reflects poorly on the author and can damage professional relationships

A better strategy is to consider **who has engaged with your work at conferences, asked insightful questions, shown interest in your research or is now publishing on it**. Early-career researchers are often excellent reviewers—they are engaged, detail-oriented, and more likely to accept review requests. Finding a postdoc in a lab that works on related topics to yours, can be a great option, this is how many associate editors find reviewers.

Regarding exclusions, this option should be used very rarely. Excluding a reviewer should be reserved for cases of **serious professional conflict** or strong evidence of bias. Simply disagreeing with someone's previous work does not justify exclusion—scientific debate is part of the process.

The peer review process is unpredictable, but by **carefully selecting**

suggested reviewers, responding to feedback thoughtfully, and maintaining professionalism, authors can increase their chances of publication while contributing to a more constructive review culture.

#09

NAVIGATING PUBLISHING CHALLENGES

Rejection is the rule, not the exception

During my PhD, I worked relentlessly on my papers, yet most were rejected multiple times. My personal record? The same paper was rejected twice in one day—first as a desk rejection in the morning, then again that evening after I reformatted and resubmitted it to a different journal. A friend of mine once received a rejection in just 35 minutes because the editor immediately considered that the paper was not a good fit. On average, my PhD papers faced at least five rejections before publication.

Rather than seeing this as failure, I viewed it as part of the process. Every rejection forced me to improve my work, and in many cases, feedback from rejected submissions helped me publish in even better journals. One of my most successful papers, eventually published in a journal with "Nature" in its title, originated from a manuscript rejected elsewhere. A reviewer from a second-tier journal had provided insightful but demanding feedback—so demanding that it exceeded the expectations of that journal (in my opinion:)). I used their critiques to refine the paper and later submitted it to a higher-impact journal, where it was accepted.

As discussed in chapter 8, rejections often happen for reasons unrelated to the manuscript's quality. Sometimes, a journal has received too many papers on the same topic in a short period. Other times, an editor may feel the paper does not fit the journal's evolving focus. These decisions can be unpredictable and frustrating. Also remember that editors and reviewers are human (for now!). We like to think of science as purely objective, but factors beyond our control —an editor and reviewer's workload, the timing of submission, even their mood—can influence decisions. The best approach? **Don't take rejection personally. Keep improving and keep submitting.**

When to appeal a rejection

Most appeals are unsuccessful because they are based on frustration rather than valid editorial errors. Journals are reluctant to overturn decisions unless a clear mistake has been made.

Some of the worst appeals I've seen come from senior researchers attempting to intimidate editors look something like this:

"I have published 400 papers and served as Editor-in-Chief of a more prestigious journal than yours [name of journal here]. Rejecting my paper is clearly a rookie mistake."

These appeals often backfire, reinforcing the perception that publishing favors the privileged. As an editor, I reject these appeals not only because they lack substance but because giving in to them would perpetuate inequality.

However, there are legitimate cases for appeal. For example:

"The main reason for rejection was low sample size. In our study system, it took three years to collect these data, and similar research in this field has been published with even smaller sample sizes. We would like the opportunity to clarify why this is a strength rather than a weakness."

A well-argued appeal that corrects a factual misunderstanding or highlights overlooked strengths can sometimes lead to reconsideration.

Similarly, appeals from underrepresented researchers who face additional barriers in publishing deserve attention:

"We are a team of researchers from Venezuela, none of whom have degrees from outside South America. We recognize that our paper lacks clarity, but we have now enlisted the help of an experienced colleague to refine it. Publishing in your journal would help increase the international visibility of our work."

In contrast, well-established researchers with 100+ publications should probably **don't appeal and just submit elsewhere**.

Some researchers, particularly those with strong egos, take an aggressive approach—calling the Editor-in-Chief directly to claim, *"You've made a huge mistake."* Interestingly, this tactic tends to be used most successfully by older male scientists, further widening the privilege gap. If you are an early-career researcher or from the Global South, **consider appealing when you believe a rejection was truly unfair.** Otherwise, move on, there are plenty of nice journals out there.

Note: sometimes there are obvious errors, such as they review the wrong file, or clear unfairnesses that deserves an appeal by anyone.

Understanding the mismatch between authors and editors

Publishing is rarely a straightforward process. Even strong research can be rejected due to factors beyond its quality—journal priorities, editorial biases, or simply bad timing.

One of the biggest challenges is the disconnect between **how authors perceive their work and how editors and reviewers see it**. Authors often view their papers as unique and groundbreaking, while editors evaluate them in the context of many similar submissions. This mismatch can cause frustration, but understanding the editorial perspective can help you navigate the system more effectively.

What you may think	What editors may think
No one reads the cover letter.	This author didn't even try to write a letter—why should we care about their paper?
Formatting is a waste of time.	This manuscript looks formatted for a different journal—was it rejected there first? Why?
I don't have time to answer all the submission questions.	These details (e.g., why this paper fits the journal, suggested reviewers) are critical for our decision- making.
This is my best work—it took years to complete.	This is the third paper we've received on this topic this week.
My figures don't need to be pretty—the data speaks for themself.	These figures look rushed and unpolished. Was the rest of the paper rushed too?
"This is the first time something did this".	Another paper overselling their results.
Extremely broad title, it is appealing to a broad audience!	Way too broad, what is the domain??

See the table for some common differences in perception.

Understanding these differences can help you better prepare your submissions. Editors and reviewers assess not only the scientific merit of your work but also its clarity, relevance, and presentation. A well-polished manuscript with clear figures and a strong cover letter will always have a better chance of success.

Rejection is inevitable—Keep moving forward

Again, rejections are part of the academic publishing process, no matter how experienced you are. The key to success is persistence. Some of the most prolific scientists have also been rejected the most.

Take Picasso or Taylor Swift—both produced hundreds (or thousands) of works, refining their craft through repetition. Science is no different. The more papers you write, revise, and submit, the better you will become at **navigating the system and communicating meaningful science.**

Rejections will come. Learn from them, improve your manuscript, and submit again. That's how you get published. **Everyone gets their paper rejected, even the researchers that you admire the most.**

#10

PUBLISHING FOR IMPACT SELECTING THE CORRECT JOURNAL

Choosing the right journal for your paper is not always straightforward. Journals constantly evolve, editorial boards shift, and reviewer feedback can be unpredictable. A manuscript that is a perfect fit today might not align with the journal's priorities a year from now. Selecting the right journal is about maximizing your chances of acceptance while ensuring your work reaches the right audience.

Finding the right fit

Clearly researches in different stages have different needs. An undergrad may need a paper to get accepted very fast before the application to grad school. An early career researcher may need a paper in a fancy journal to get a job. A senior researcher may want to maximize readers or may want to influence some policies. A common mistake is assuming that the best journal is simply the one with the highest impact factor. While impact factors can enhance visibility, the most important factor is whether the journal's readership aligns with your research. A well-placed paper in a specialized journal may have a far greater impact within a field than a generalist journal with a broader audience but little interest in your topic. Looking at where similar studies have been published and seeking advice from colleagues and mentors can provide valuable guidance. Also impact factors can change a lot in a few years. Do your research on where you want to submit, and always have a plan B and C just in case it gets rejected.

Framing your research for the right audience

How you present your study influences where it will be published. The same dataset can be framed as a local case study or as a test of a broad theoretical concept. A manuscript focused on a specific species and location may be best suited for a regional journal, whereas emphasizing general principles can make it more appealing to an international readership.

Titles, abstracts, and introductions play a crucial role in this framing. A local study often names species and locations explicitly, while a more theoretical approach might avoid specifics until the methods section, focusing instead on broader scientific relevance. Neither approach is inherently better, but if the goal is publication in an international journal, the framing should reflect that.

When to publish in your local language

Publishing in English is the norm in many disciplines, but it is not always the best choice. Research aimed at policymakers, local stakeholders, or public health officials can have greater impact if published in a language they are comfortable reading. Many environmental policies, for example, rely heavily on local-language literature. If the goal is to influence practice rather than maximize academic citations, publishing outside English-language journals can be a strategic decision.

Balancing Prestige and Practicality

The key to effective publishing is balancing ambition with pragmatism. Targeting high-impact journals is worthwhile, but spending years reformatting a paper for rejection after rejection might not be the best use of your time and can get frustrating. At the same time, settling for a journal with little relevance to your work diminishes its reach. The best approach is to understand the field, frame the paper appropriately, and be willing to adapt. Talk to peers and more senior researchers about this. Researchers spend a lot of time thinking about publishing strategies!

Type of publications

Scientific writing takes different forms depending on the purpose and audience. Understanding these different formats helps researchers choose the best outlet for their work or ideas.

Research papers

Research papers present original findings based on experiments, observations, or data analysis from databases. These are the most common type of scientific publication and typically follow the "IMRaD" structure: Introduction, Methods, Results, and Discussion. Research papers contribute new knowledge to a field, whether through novel discoveries, methodological advancements, or theoretical developments. They undergo peer review to ensure scientific rigor before publication. **These are the typical papers that come up from PhD thesis and are the building blocks of science.**

Review papers

Review papers summarize and synthesize existing research on a specific topic, providing a comprehensive overview of current knowledge, trends, and gaps. They do not present new experimental data but instead analyze and interpret findings from multiple studies. **Most journals appreciate these papers since they can be highly cited.** Some are narrative, while others use systematic or meta-analysis approaches to draw conclusions from large datasets. I like to think that the first chapter of every PhD thesis can be a review paper that can be published. They can also be very cheap to produce, since there is no field or lab work involved.

Opinions and replies

These articles provide perspectives, critiques, or commentary on scientific issues, policies, or recent publications. Opinion pieces express personal or expert viewpoints on a controversial or emerging topic. Replies are responses, critiques or discussions on previously published work. Unlike research papers, these are typically shorter, less data-driven, and may not require peer review. They are much faster to write, and can have high impact, but may not count so much towards CV building as reviews or research papers. They can be really fun to write and can give the authors a lot of visibility.

Preprints

Preprints are **early versions of research papers** shared on open-access platforms **before peer review**. They allow rapid dissemination of findings, enabling feedback from the scientific community and increasing visibility. While preprints can speed up knowledge sharing, they lack formal peer review and should be interpreted cautiously. Most journals now accept submissions that were first posted as preprints, and some journals even look at them to invite them for submission.

Book Chapters

Book chapters are contributions to e**dited academic books** that focus on specific aspects of a broader topic. Unlike journal articles, book chapters **may not be peer-reviewed as rigorously** but are often invited contributions by experts. They can include reviews, original research, or theoretical discussions and provide in-depth analysis that may not fit within a standard journal format. Book chapters can be great places to summarise knowledge or to come up with new ideas but are typically cited less frequently than journal articles since books are harder to come around than papers, and may not be accessible in many databases. Some can be really influential and prestigious, but many receive less attention than similar text published in a journal. For early career researchers they typically are less effective at career building than papers in journals.

Predatory journals

A **predatory journal** is a deceptive academic publication that **lacks legitimate peer review** and **exists primarily to collect fees** from authors. These journals **exploit researchers**—especially early-career scientists and those unfamiliar with the publishing process—by **promising fast publication** with little to no editorial oversight.

Unlike reputable open-access journals, which charge **Article Processing Charges (APCs)** to make a profit but also to maintain editorial quality, predatory journals **prioritize profit over integrity**. They often **mimic well-known journals**, fabricate impact factors, and **mislead authors** by falsely claiming rigorous review processes.

The rise of predatory journals threatens scientific credibility, making it harder to distinguish genuine research from bad-quality publications. Thousands of such journals exist, and their numbers are growing. Researchers should be cautious of unsolicited invitations to publish, especially when a journal:

* Seems too good to be true: they claim to know your work and give you a great discount due to the quality of your science (those emails can be really good!).

* Lacks a **recognized editorial board**.

* Is **not indexed** in Web of Science, Scopus or the Directory of Open Access Journals (DOAJ).

* **Falsely claims indexation** in reputable databases, lists impact factors that don't exist, or references questionable indexing services.

It is not always easy to detect predatory journals. When in doubt, always **verify a journal's credibility** before submitting your work to avoid the trap of predatory publishing. **Publishing in a predatory journal can damage your reputation** (raising questions about why you chose that outlet) and **make it difficult or even impossible to publish the work elsewhere**. Be cautious—**don't be fooled by flattering emails or by journals with impressive-sounding names**.

Open access vs. traditional publishing

Scientific publishing operates under two main models: **open access (OA)** and **traditional publishing**. Each has distinct advantages and drawbacks, mainly revolving around **who pays and who has access to the published research**.

Open access (OA): pay to publish

In the **open access model**, research is **freely available to anyone without subscription fees**. However, the cost of publishing is shifted to the authors, who must pay **Article Processing Charges (APCs)** to cover editorial and production expenses.

Pros:

* Makes research **accessible to everyone**, including scientists in low-resource settings.

* Increases **visibility and citations**, as paywalls limit readership.

* Complies with many f**unding agency requirements** for publicly funded research.

Cons:

* **High APCs** can be a barrier, especially for researchers without funding.

* Some **predatory journals** exploit the OA model by charging fees without providing proper peer review.

* Not all **institutions or grants cover OA costs**, making it an expensive option for some researchers.

Traditional publishing: paywall restrictions

In the **traditional model**, authors typically do not pay to publish, but **readers must pay** to access the content through **subscriptions or individual article fees**. Journals generate revenue from institutional subscriptions, often making content inaccessible to those without access through a university or organization.

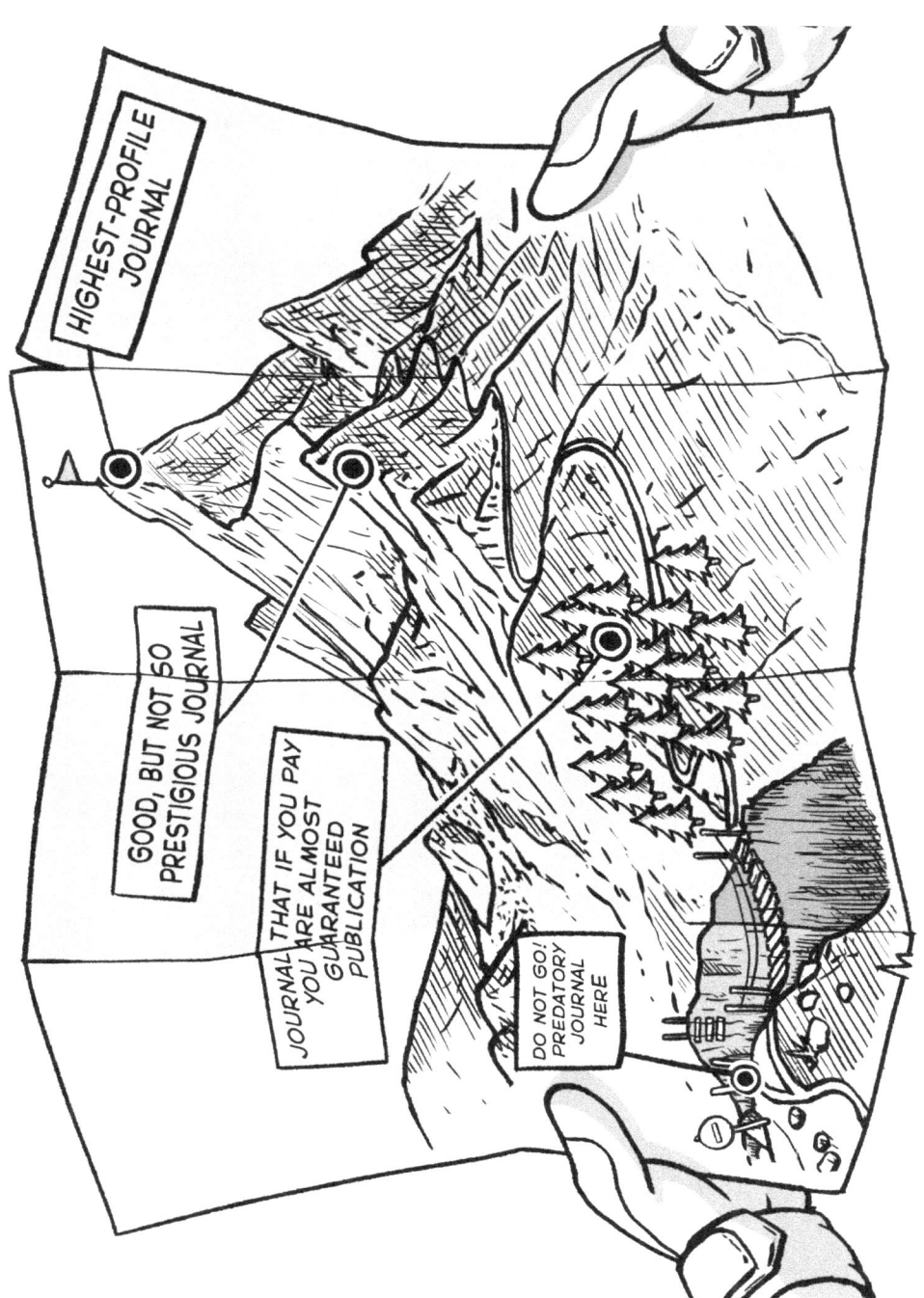

Pros:

* No **direct cost** to authors.

* Many **high-impact journals** still operate under this model.

* Often perceived as having **rigorous editorial standards** (though OA journals can be just as selective).

Cons:

* Research is **locked behind paywalls**, limiting accessibility.

* Institutions must **pay high subscription fees**, restricting access for underfunded universities.

* Authors might not have full access to their own work (e.g., to even share it with students in a class) if they lack institutional subscriptions.

Which model is better?

The **best model depends on priorities and the available journals**. Sometimes the best journal for your paper is open access, sometimes it is not. The landscape of scientific publishing is evolving, with **many funders now requiring OA**, pushing the system toward greater accessibility for readers. However, costs remain a significant challenge, especially since OA has now added a new job for many researchers: finding funds for publishing. The debate between OA and traditional publishing is far from settled, but **it looks like in the future all journals may be OA.**

If you really want to publish in an OA journal and do not have the funds, do not panic. **You can always request a waiver for the cost (APCs).** I should say that if you explain your reasons well, it is likely that you will get a waiver. In my experience, societal journals are especially good at giving waivers.

#11

PLAGIARISM AND ACADEMIC INTEGRITY

Plagiarism is one of the most serious ethical violations in academic publishing, yet its perception and consequences vary widely across cultures. In some regions, borrowing text or ideas without explicit citation is not seen as so bad. I remember in Argentina, one of my first professors told us that a way to deal with English grammar was to just copy full sentences from well written papers. I can really relate to the struggles of non-native speakers, but that is plagiarism... In the academic communities of the United States and Europe, plagiarism is considered a major offense—one that can lead to retraction of papers, damaged reputations, and even career- ending consequences.

What is plagiarism?

At its core, plagiarism is using someone else's words, ideas, or work without acknowledging them. This does not just apply to copying large sections of text—self-plagiarism, paraphrasing without citation, and failing to attribute key concepts can all fall under this category. Many researchers engage in unintentional plagiarism, often due to poor citation habits or a lack of familiarity with best practices in academic writing.

Plagiarism can take many forms, ranging from direct copying of sentences from another source to patchwriting, where a writer slightly alters a passage without truly making it original. Self- plagiarism, or reusing significant portions of one's own previously published work without proper disclosure, is also considered problematic, as it misleads readers and journals about the novelty of the research. I personally don't consider Self-plagiarism as a type of unethical practice, but many people do. Also typically we sign away our rights to our own text when we publish in many journals. So that text belongs to someone else (weird..)

Why is plagiarism taken so seriously?

In academic publishing, originality and attribution of contributions are fundamental principles. Knowledge builds on previous work, but researchers are expected to contribute new insights while properly crediting past scholarship. Most institutions emphasize strict ethical guidelines to protect intellectual property and ensure fairness in academic recognition.

In contrast, some cultures historically value the repetition and transmission of established knowledge rather than emphasizing individual contributions. In such contexts, reusing existing text without citation may not carry the same negative connotation. However, for anyone working in international academia— particularly in the U.S. and Europe—it is crucial to follow their ethical standards to avoid serious professional consequences.

How to avoid plagiarism

The best way to prevent plagiarism is to develop strong citation habits and learn how to properly integrate sources into your writing. Here are a few key strategies:

1. **Use Proper Paraphrasing.** Paraphrasing does not mean just changing a few words while keeping the structure of the original sentence. A proper paraphrase reframes the idea in your own

words and style while still crediting the original source. If the wording is too similar to the original, it is still plagiarism. As someone told me "the words need to come from your brain".

2. Avoid Self-Plagiarism. If reusing parts of your own previously published work, disclose it clearly and seek permission if necessary. Many journals require authors to state whether any sections of their manuscript have been published elsewhere.

3. Use Plagiarism Detection Tools. Many journals use software such as Turnitin or iThenticate to detect copied text. Authors can also run their manuscripts through these tools before submission to catch any accidental plagiarism.

Consequences of plagiarism

The consequences of plagiarism can be severe. Journals may reject manuscripts outright, retract published papers, or ban authors from submitting future work. They may even contact your institution. Plagiarism scandals have caused researchers to lose their jobs, funding, and reputations. Even unintentional plagiarism is treated seriously in many academic institutions, so it is always better to err on the side of caution. This of course adds another complication to non-native English speakers that may need to rephrase their own words to make them acceptable.

In academia, integrity is as important as the research itself. Science is based on trust. While the definition and perception of plagiarism may vary across cultures, researchers working in international scientific communities must respect the strict ethical standards of academic publishing. When avoiding plagiarism, researchers not only protect themselves from misconduct allegations but also contribute to a culture of transparency and intellectual honesty.

#12

ARTIFICIAL INTELLIGENCE (AI) IN SCIENTIFIC WRITING

Resistance is futile

Artificial Intelligence (AI) is becoming increasingly important in science, and its role is only expected to grow. Whether we like it or not, AI is here to stay. Many of us anticipate that it will play a transformative role in research, including writing and publishing. AI tools are already changing the way scientists draft manuscripts, refine language, format text, analyze data, and even formulate research questions.

There is some resistance to the use of AI, and we need to be aware of the different requirements. Some publishers strictly forbid the use of AI, while others don't seem to care. And some, arguably in the middle, now ask for statements where you need to be explicit if you use it or not. I do not even name the journals since I expect that this will change very quickly for a more AI friendly perspective.

Resistance is futile. In scientific writing, AI is already deeply embedded in the publication process. Many publishers now offer AI-assisted grammar and formatting tools to help authors improve their manuscripts before submission. Soon, we may see **standardized AI**

assistants that go beyond fixing typos and actively help in structuring arguments, summarizing papers, or even generating first drafts of sections.

For **non-native English speakers**, this is a true blessing. A friend of mine recently told me, *"I finally sound smart in English."* AI tools can help level the playing field, reducing the disadvantage that many researchers face when publishing in English-language journals.

Please note that for researchers just beginning to write, relying heavily on AI can be risky. If you're still learning what good scientific writing looks like, **you may not yet have the judgment needed to evaluate or revise AI-generated content effectively**. I recommend a limited use of AI tools at first—perhaps for grammar correction or translation—but focusing instead on building a solid understanding of structure, style, and clarity. Once you have more experience and confidence, AI can become a powerful ally in refining your work.

Great ways of using AI in your daily writing

AI can be a **powerful assistant** in scientific writing, but like any tool, it works best when used strategically. Here are some of the most effective ways to integrate AI into your daily writing routine:

1. Refining and polishing your writing

AI can significantly enhance clarity, conciseness, and readability. You can write a rough draft and ask AI to **improve sentence structure, enhance fluency, and fix grammatical errors**. This is especially helpful for non-native English speakers who want their writing to sound more natural and polished. However, while AI can smoothen your text, it is not always perfect, so manual review is necessary to ensure precision.

2. Rewriting and formatting

Need to restructure a paragraph? AI can help by **suggesting alternative phrasings, adjusting sentence order, and reformatting text** to improve readability. Additionally, if you need to reformat a manuscript to fit a specific journal's style, AI can help tweak citation styles and layout requirements—though final checks are always essential.

3. Summarizing text

AI is great for **summarizing papers, articles, or even entire books**. If you are reviewing a long paper, you can use AI to generate a **concise summary of the main findings**. This can speed up literature reviews and help you quickly determine whether a paper is relevant to your research. However, there is no substitute for reading a paper yourself, as AI can sometimes misrepresent complex arguments.

4. Reducing word count

Journals often impose strict word limits, and AI can assist by **cutting unnecessary words** while retaining meaning. You can ask AI to make your text **more concise, eliminate redundancy, or simplify overly complex sentences**—a valuable tool when struggling to fit within a word limit without losing essential content.

5. Generating writing ideas and overcoming writer's block

Staring at a blank page? AI can help **brainstorm ideas, suggest paper structures, or even generate first drafts of specific sections**, if given the right prompts. While AI cannot replace original thinking, it can **jumpstart the writing process** by providing outlines, possible phrasing, or alternative ways to express your argument.

6. Editing references and citation management

One of the **most tedious** aspects of scientific writing is **formatting references**. AI can help identify **missing references, suggest proper citation styles, and even generate citation summaries**. However, AI often **hallucinates references**, meaning it can fabricate non-existent citations—so every reference should be double-checked against actual sources.

I like to think of AI as a **very eager 15-year-old intern— enthusiastic, hardworking, but often clueless**. You cannot trust it to do everything on its own, but for tedious, repetitive tasks (like **editing references**), it can be a fantastic time-saver. AI is not a replacement for **critical thinking, creativity, or scientific expertise**, but when used wisely, it can significantly improve efficiency and writing quality. Again **editing is easier than creating, so let AI do an outline that we can work on!**

The challenge of using AI in science

However, scientific writing is not just about producing text—it is about generating **new knowledge.** This is where AI presents a fundamental challenge. By design, AI systems work by analyzing and synthesizing **existing information**, meaning that they are inherently limited by what has already been published. They cannot create new discoveries, only **rearrange, summarize, and predict based on past data.**

Science is about pushing boundaries, questioning assumptions, and finding new patterns. Relying too heavily on AI for scientific writing could lead to work that is **predictable, derivative, or even misleading.** A machine learning model does not understand the nuances of a novel hypothesis or the complexities of a cutting-edge experiment—it only generates text that is statistically probable based

on existing literature. This is why scientists must remain in control of their writing and ensure that AI is used as a tool rather than a substitute for original thinking.

Another critical challenge is **accuracy.** AI-generated text is prone to **hallucinations**—false or misleading statements that sound plausible but have no factual basis. This is especially dangerous in scientific writing, where precision is essential. If AI-generated summaries or explanations are used without careful fact-checking, they could introduce errors that compromise the integrity of research. This can deeply affect your reputation, so do not make this mistake.

I expect that if I revisit this chapter in three years, my perspective may have evolved. AI is advancing rapidly, and its capabilities will likely expand in ways we cannot yet predict. For now, my advice is simple: **stay informed, embrace AI where it adds value, but always use it responsibly and do not assume accuracy.**

Crafting effective prompts for academic research in ChatGPT

The quality of AI, such as ChatGPT's responses, depends on how well you frame your prompts. A well-structured prompt provides clarity, context, and direction, ensuring relevant and useful results.

How to write good prompts:

Be clear and specific: ChatGPT responds best to well-defined questions. Instead of a vague prompt like *"Tell me about invasive species,"* which could generate an overly broad and generic response, a more precise prompt such as *"Summarize key ecological impacts of invasive pines in South America"* helps narrow the focus and provides more relevant insights. Specificity ensures that the response aligns closely with your research needs.

Define the format: Clearly stating the desired format helps structure the response. If you need a list, a summary, or a comparison, say so explicitly. For example, instead of just asking *"How do invasive plants alter soil chemistry?"*, a better prompt would be *"List three mechanisms by which invasive plants alter soil chemistry, with examples and references from scientific papers."* This instructs ChatGPT on how to organize the information, making the output more useful.

Provide context: ChatGPT does not inherently know your specific focus, so adding context improves the response quality. If your research involves a particular aspect of a broader topic, specifying details like *"Explain how allelopathy contributes to pine invasions, with examples from recent studies"* ensures that the response is relevant to your area of interest. Without context, you may receive a general explanation that lacks depth. You can also upload a PDF or a web link of some specific paper to help guide it.

Ask for refinements: If the first response is too general or lacks depth, you can fine-tune it by requesting adjustments. Follow-up prompts like *"Make this more concise,"* or *"Expand on point three with examples from recent literature"* allow you to control the level of detail. This iterative process helps refine the information until it meets your needs.

What to avoid:

Overly broad questions: If your question is too general, the response will likely be vague and unfocused. For example, *"Tell me about cancer"* is too broad and will generate a general overview rather than a meaningful academic insight.

Multiple questions in one prompt: Asking multiple unrelated questions in a single prompt can confuse the model and result in a disorganized response that is more shallow. Instead of asking *"What are the impacts of invasive pines, and how do they compare to eucalyptus?"*, it's better to split it into two separate queries to get more structured answers. You will get more machine power into each question if you make multiple questions and not one big one.

Assuming accuracy: AI generates responses based on patterns in its training data but does not verify sources. It may provide plausible but incorrect or outdated information. Always cross-check key facts and verify references with primary literature to ensure accuracy.

Well-crafted prompts make AI a powerful tool for summarizing, brainstorming, and refining research ideas. By structuring queries thoughtfully and refining responses iteratively, researchers can maximize its usefulness while maintaining academic rigor.

AI tools for scientific writing

There are already several AI-powered tools that can assist researchers in writing, editing, and even summarizing scientific literature. Some of the most useful for me (and that I have tried) include:

1. ChatGPT/DeepSeek/Claude/Gemini and other general AI assistants

ChatGPT, DeepSeek, Claude, and Gemini are among the most well-known AI writing assistants. They can help generate text, summarize papers, suggest improvements, and refine arguments. While incredibly useful for brainstorming and drafting, it is important to remember that they do not "understand" science—they simply predict likely sequences of words. This means that any AI-generated text should be carefully reviewed and fact-checked before use in a scientific paper.

2. Consensus.app

Consensus.app is an AI-powered search engine that helps researchers find relevant scientific literature quickly. Instead of returning a long

list of papers like Google Scholar, it synthesizes key insights from multiple studies, providing a more direct summary of the existing research on a given topic. This can be useful when conducting literature reviews or identifying key trends in a field.

3. ChatPDF

ChatPDF is an AI-powered tool that allows researchers to interact with PDF documents, including research papers. Instead of manually skimming through long papers, users can ask specific questions about a document, and ChatPDF will generate responses based on its content. This can be especially useful for reviewing large amounts of literature quickly, extracting key findings, or clarifying complex sections of a paper. While highly efficient, it should be used as a complement to deep reading, not a replacement for critical analysis.

4. Perplexity.ai

Perplexity.ai is an AI-powered research assistant that helps researchers find and summarize scientific literature efficiently. Unlike traditional search engines, it provides concise summaries with direct citations, reducing the risk of fabricated references. It is particularly useful for literature reviews, fact-checking, and exploring new topics by retrieving relevant academic sources.

5. DeepL

DeepL is one of the most accurate AI-powered translation tools available, making it particularly useful for researchers who work in multiple languages. It can help translate research papers, abstracts, or scientific terminology while maintaining nuanced meaning. While DeepL outperforms many other machine translation tools in accuracy, complex scientific concepts may still require manual adjustments and proofreading to ensure clarity.

6. Grammarly

Grammarly is an AI-driven writing assistant that helps researchers refine their writing by correcting grammar, spelling, punctuation, and style. It also offers suggestions for clarity and conciseness, making it useful for improving academic manuscripts, grant proposals, and emails. While helpful, Grammarly does not always recognize the nuances of technical scientific writing, so its suggestions should be reviewed critically.

7. Writefull

Writefull is an AI-powered tool designed specifically for academic writing. It helps researchers by suggesting word choices, checking grammar, and analyzing sentence structures against a vast database of published academic papers. It also offers a language feedback feature tailored to formal writing. Writefull is particularly valuable for non-native English speakers or those looking to improve the academic tone of their manuscripts. We use this at the British Ecological Society and it is very useful.

8. ResearchRabbit

ResearchRabbit is an AI-powered literature discovery tool that helps researchers find relevant papers and build connections between studies. Unlike traditional search engines, it creates visual networks of related research, making it easier to track developments in a field. This is especially useful for identifying influential papers, exploring citations, and discovering emerging trends. While powerful, it should be used alongside conventional literature review methods to ensure comprehensive coverage of a topic.

These tools can make scientific writing more efficient, especially for early-career researchers or those writing in a second language.

However, they should never replace critical thinking, originality, or rigorous peer review. AI can assist in writing, but it cannot (and should not) generate entire research papers without human oversight.

AI is already shaping the way we write, edit, and publish scientific research, and its influence will only grow. Used wisely, it can **enhance productivity, improve language clarity, and speed up the writing process**—but it also comes with risks. Over-reliance on AI could lead to **unoriginal, inaccurate, or misleading writing**, which goes against the very nature of science.

One growing concern is that some AI tools can generate plagiarized content—copying published material verbatim or slightly paraphrasing comprehensive synthesis without proper attribution (see previous section on plagiarism). This not only violates academic integrity but may also result in serious ethical and legal consequences for researchers.

For researchers, the key is to **stay informed, experiment with AI tools, and use them responsibly, following rules (e.g., do not upload unpublished work of others without a clear agreement) and local laws.** AI should be a **tool to assist scientific writing, not a substitute for human reasoning and creativity.** The best science will always come from **real human insight, innovation, and critical thinking (or I hope so!).**

#13

AFTER PUBLISHING – ADVERTISING YOUR WORK

One of the main reasons we publish is to ensure that people—colleagues, potential employers, lawmakers, and practitioners—know about our research. Ideally, once a paper is published, all researchers working on related topics would automatically receive it, read it, and consider its implications. However, we do not live in an ideal world. Tens of thousands of papers are published each year, and many (most?) researchers struggle to keep up with the literature. Some rely on social media, others on personal recommendations from colleagues, and many simply do not make a systematic effort to stay updated.

If you want your work to have an impact, **you need to actively promote it. This is part of your job!** This is particularly important for early-career researchers (ECRs), who are not yet well known in their fields and are therefore less likely to attract attention when their papers are published. Advertising your work can take many forms, from conference talks and emails to social media posts and institutional press releases. While self-promotion may feel uncomfortable, it is a valuable skill that ensures your research reaches the right audience.

If you're unwilling to promote your own work, why should others do it for you? In some cultures—such as in the United States—self-promotion is widely accepted. In others, it may be perceived as bragging. But regardless of cultural norms, the reality is that we live in a world overwhelmed by information. If we want our research to be noticed, we must make it easier for others to find and engage with it. That means stepping outside of our comfort zones and finding respectful, authentic ways to share our work. Help others discover the value of your research—because if you don't, it may never reach the audience it deserves.

Presenting your work at conferences

Giving a talk at a conference is one of the best ways to share your research with an engaged audience. In many disciplines, it is perfectly acceptable to present published papers as a main part of your presentation, and it is great to mention your work when talking about not yet published work. While conferences are often seen as venues for showcasing new research, discussing published work allows for deeper engagement, as the audience can access the full paper afterward. This can lead to **new collaborations, citations, and invitations for future talks.**

When presenting published work, it is helpful to emphasize its significance and relevance to ongoing discussions in the field. A well-delivered talk can generate interest beyond those who would have discovered the paper on their own.

Contact experts who may be interested

Another effective way to promote your work is by directly reaching out to researchers who might find it relevant. This can include both

people you know and those you do not. Sending an email or Direct Messenger (DMs) in social media may feel intrusive, **but many scientists appreciate receiving papers that align with their research interests.** This is especially important now, that so much is being published and it is so easy to stay behind in the literature.

If you are an ECR, this can also be an opportunity to introduce yourself to senior researchers in your field. A well-crafted email can foster connections, open doors to collaborations, and even lead to invitations for talks or discussions.

For example, a simple and respectful email might look like this:

Subject: *Recent Publication on* [Your Topic]

Dear Dr. Perez,

I recently published a paper on [topic], *and I know that you have an interest in this area of research. I am a PhD student working on* [related topic], *and I would love to hear any thoughts you might have on our findings—of course, only if you have the time. Here is the link to the paper/PDf:* [Insert Link/ attach PDF].

Thank you for your time, and I appreciate any feedback you may have.

Best regards, [Your Name]

A short, polite message like this is unlikely to be seen as spam, especially if the recipient has a genuine interest in the topic. Even if they do not respond, they may still read the paper or remember your name when they encounter your work in the future.

Using social media to promote your work

Social media has become an essential tool for academic networking and research dissemination. Platforms like Twitter (X), LinkedIn, Facebook, and BlueSky allow researchers to share their work with a broader audience.

Some researchers feel uncomfortable posting about their publications, worrying that it may come across as self-congratulatory. However, **advertising your work does not have to feel like bragging**—it can be framed as sharing knowledge with the community.

For example, instead of simply stating, *"I just published a new paper!"* you might post something more engaging, such as:

> *"At one point, I thought this paper would never see the light of day. But after years of work, we finally published our findings: Singing to plants increases seed production! Take a look if you're interested. Huge thanks to Pedro Pascal and John Wilson for their support throughout this project!"*

This type of post not only promotes your paper but also makes it more relatable to other researchers, especially those struggling through long review processes.

Another good practice is to promote **others' work alongside your own**. If you regularly share interesting papers from colleagues, people will be more receptive when you post about your own research. Building a habit of engaging with other researchers' work— commenting on their posts, retweeting their findings, and participating in discussions—can help create a supportive academic community.

Communicating outside of your discipline

Making Your Work Accessible and Visible Is Part of Your Job!

Publishing a paper is only the first step in ensuring that your research has an impact. Actively promoting your work is not about vanity—it is about making sure your findings reach the right audience. Ensuring that your research reaches beyond your immediate academic circle is crucial for making a real impact. Whether your work has direct applications or is focused on basic science, there is always value in communicating its significance to society.

Engaging with the public serves multiple purposes:

*   It raises awareness of important scientific advancements.

*   It helps inspire future generations of scientists.

*   It demonstrates the value of research to taxpayers, who ultimately fund much of our work.

By communicating effectively, we bridge the gap between science and society, ensuring that research is not confined to journal paywalls but contributes to public understanding and progress.

Using your institution's public relations (PR) department

Most universities and research institutions have media or public relations (PR) teams dedicated to promoting researchers and student achievements. These teams are often eager to share research updates but rely on researchers reaching out with newsworthy findings.

If your work has potential public appeal—whether due to groundbreaking discoveries, policy relevance, or local interest—it is worth contacting your PR department. Many researchers assume that only papers published in high-impact journals like **Science** or **Nature** are newsworthy, but that is not always the case. A study with a

compelling story, societal relevance, or practical applications may be just as valuable for media outreach.

A simple email to your university's PR team can significantly boost your research's visibility. In many cases, they can:

* Draft and distribute press releases.

* Arrange interviews with journalists or media outlets.

* Promote your work through university websites and social media channels.

Navigating media engagement: a word of caution

While working with journalists or university press offices can amplify your research's reach, it is important to maintain accuracy and control over how your work is represented. Media coverage can sometimes:

* Oversimplify complex findings.

* Exaggerate claims for broader appeal.

* Omit key details or fail to acknowledge collaborators.

To avoid misrepresentation, always ask for final approval of any press release or news article before publication. While most journalists mean well, they may unintentionally distort findings in an effort to make them more digestible to the public. Reviewing the final version ensures that your research is accurately communicated and prevents misunderstandings or unnecessary conflicts.

By strategically engaging with both academic and public audiences, you can maximize the impact of your research, ensuring it is not only published but also that it has an impact.

#14

THE FUTURE OF SCIENTIFIC PUBLISHING

One of the most fascinating aspects of scientific publishing is that it is constantly evolving. Over the past two decades, we have witnessed dramatic shifts—from the rise of online publishing to the growing dominance of open-access models and, more recently, the increasing role of artificial intelligence in research and writing. These transformations are far from over. Scientific publishing will continue to change, sometimes in ways we can predict, but often in ways we cannot.

Keeping an eye on these changes is more important than ever. Journals that are prestigious today may lose their influence in a few years, while new platforms and alternative models could rise in importance. Preprints, which were once seen as informal and unvalidated, are now gaining visibility, and their role in disseminating research may expand further. Peer review, traditionally an unpaid and voluntary effort, might shift toward paid models or AI-assisted reviews, knowing how hard it is to find reviewers in most journals. The way scientific contributions are evaluated—through journal impact factors, citation indexes, or alternative metrics—may also undergo significant revisions as academic institutions rethink what truly matters in assessing research impact.

It is even possible to imagine a future where AI plays a central role in scientific writing. Perhaps researchers will input hypotheses and data, and AI-driven systems will generate full manuscripts, automatically formatting and structuring papers according to journal requirements. While this may seem far-fetched, the pace of technological advancement suggests that the fundamentals of how we communicate science may change faster than we expect.

What is certain is that change is coming, and those who stay informed will be better prepared to navigate the evolving landscape of scientific publishing. Being adaptable—whether in choosing where to publish, understanding new models of peer review, or embracing emerging technologies—will be essential for researchers in the years ahead.

ESSAYS ON PUBLISHING

For this section, I invited colleagues from around the world—researchers and editors I deeply admire for their clarity, productivity, and thoughtful approach to scientific writing—to share their best advice. These are people with extensive experience publishing and editing papers, and I wanted you to hear directly from them: what strategies or tools have helped them the most? What obstacles do they see researchers facing, and how do they suggest overcoming them?

So I asked a simple question:

"Can you write a short essay offering advice to researchers on how to write scientific papers? What tool or strategy has been most helpful in your experience? Feel free to discuss the main challenges you've encountered—and how you've addressed them."

I really enjoyed reading their responses, and I learned quite a few new things myself. That's one of the great joys of doing science: we never stop learning—not even when it comes to writing and publishing.

I hope these essays will be as helpful and inspiring for you as they were for me. As with everything in this book, you might agree with some ideas more than others, but simply engaging with these diverse perspectives is valuable. There's always something to learn.

Matthias C. Rillig

Soil Scientist

The importance of establishing a writing routine

In most researchers' minds, and I literally hear this all the time, writing and the act of 'doing research' in the lab or the field are two different things. Doctoral students first do a lot of research, and then they 'write things up' towards the end of their degrees. Similar observations are also valid for many postdocs. Some people claim that they don't like writing, or that they feel they are no good at it, even though they consider themselves competent researchers. This further exacerbates the problem of not writing for long periods of time (like months), and then to expect it to all come together once the writing starts.

The fundamentally important insight is that as a researcher you are an author, by necessity. These are not separate aspects. A key step towards becoming a good writer – as for anything – is to practice writing, and this is best achieved by establishing a writing routine. When people join my lab I always tell them: you should always be writing. If you don't have results yet: work on material for the introduction of a potential paper. If you are getting results, start writing them up and work on the discussion. Write text that could be used in an opinion paper. Write material that could be used in a fellowship or grant application later. And write in English, and in your own words, don't copy anything, ever; always paraphrase things in your own words. Read and observe how others have written their papers. Take a note of what you liked about others' writing.

You are the only person who knows what kind of writing routine works for you. Do you work on your own, or do you enjoy the accountability that comes with working in a group setting (like a writing retreat you could schedule within your department)? Do you like to write in the morning, or later in the afternoon? Do you listen to music, or do you like white noise, like in a café, or do you require absolute quiet? How do you get into the writing 'mood', what little routine do you follow, like making some tea of coffee, cleaning your desk off, or lighting a candle? At what intervals do you write, every day, twice every week? Do you find that you can write as things pour out of you, or do you need a clear outline or a plan first? You can find out by experimenting with what works for you, nobody knows until they try. The important thing is that you do establish a habit of always writing.

When you always and consistently write something, you will become better at it over time. Also, some people do get ideas while they are writing (I do, for example), whereas others don't (do you know in what group you are?). If you are in the group of people who do get ideas while writing, then this is also a way to become more creative. And that is an added bonus.

David Wardle
Ecologist

Here are ten points that continue to guide my own scientific writing. These are of course what works for me and are loaded with my opinions – others might have very different ways of doing things, and as always there are multiple equally effective alternative routes to the end goal.

1. Every manuscript needs a really simple take home message – what it is that you have discovered and why it is important,

2. ideally something that can be stated in a single sentence. Everything in the manuscript, from your hypotheses to your figures and tables, should then somehow connect to it.

3. Following from the first point and to help conciseness, every sentence in the manuscript should do something useful - which begets the need to carefully check that every sentence actually is needed for developing the story. This is the scientific writing equivalent of the novelist Kurt Vonnegut's advice about creative writing: 'Every sentence must do one of two things— reveal character or advance the action'.

4. Many folk learning the process encounter real difficulty in structuring consecutive paragraphs in a logical way, particularly the Introduction and Discussion. What I still always do is first construct a road map listing all the paragraphs, with a few bullet points for each paragraph, before I even begin writing. I'm surprised that more people don't do that.

5. If you want people to read and understand your papers then it is important to keep things simple and easy to follow. This applies not just to text, figures and tables but also the choice of statistics – too many folk use unnecessarily complex statistics when simpler and easier to understand statistics would also work just as well. Remember that your goal is to clarify not impress.

6. Remember to cite all relevant literature and ensure that you do not miss anything important – those whose work you have failed to cite might be your reviewers or editors. And try not to overlook literature from particular regions or groups -- note that there are lots of excellent papers from outside of places like the US and UK that merit citation even if they haven't had the publicity.

7. Very carefully check and recheck all your data analyses and figures and tables – too many good and honest folk end up having to publish embarrassing corrections or even retractions because of some mistake that changed the conclusions and that was not discovered until after the paper was published.

8. Avoid the temptation to salami slice work into 'minimum publishable units (MPUs)' -- quality always trumps quantity in the long run. There are increasing numbers of scientists authoring 50 or more papers in a single year and I struggle to see the satisfaction in that – it is far more rewarding to focus on producing truly meaningful papers that one has a personal investment in. Arguably quality trumping quantity applies to all parts of the scientific enterprise, ranging from research group size to grant income.

9. Always make sure that your manuscript is in the style and

format of your target journal and not some other journal, otherwise reviewers and editors might conclude that it has previously been rejected by that journal which is not the impression that you want to give.

10. Be realistic in your journal choice. Many folk aim unrealistically high and then endure a lengthy chain of disappointments until it eventually cascades to the sort of journal that it should have been submitted to in the first place. Seven years of screening Science submissions while on their Board of Reviewing Editors left me thinking that about half the papers sent to Science should probably not have been sent there.

Getting a paper accepted in any journal is no easy task especially when you are starting out – it can be a long and slow journey, and it usually takes longer than you plan, but it is incredibly rewarding when you do get there, and tenacity always pays off.

Ana Pineda
Academic Writing and Productivity Coach

When Martin asked me to write this piece first I jumped of excitement. Then I panicked wondering from all the things I speak about, how could I choose one tip to improve your scientific writing.

And lastly, I procrastinated till the last moment. Now, here comes the tip:

Writing is not editing. Nor reading.

Let me explain it. When we finally find time to sit to write, most of us have built an ineffective habit since childhood. Picture this, you write one sentence, then read it, and correct a few words. You continue and write the next sentence, you read that one and the one before. Correct some words and think you're doing it terribly and that this doesn't make sense.

But you continue writing and suddenly you wonder in which paper you read it to insert the citation. And you cannot remember, so while you start thinking you don't know enough about the topic, google there we go. Luckily you find a couple of papers, print them to fully read them, and now you're there at the printer chatting with a colleague sipping the dirty coffee from the machine when this was supposed to be a writing session.

Or if you don't like coffee, you may be reading the paper you just printed. When this was supposed to be a writing session.

This scene is something I and most (if not all) of my students have lived. And a pure example of multitasking.

Multitasking is not productive. And why multitasking is so bad for you write comes from different aspects:

1.   Task switching needs brain power, so you'll be slower when you multitask than when you do the tasks at separate moments.

2.   Writing is a creative task- we need a part of our brain that brings ideas and puts them in words in your mind and then in the document. It's a task that needs flow!

3.   Editing is a task that requires critical thinking and works with a different part of your brain. The one searching for mistakes and what is not working. And not only interrupts your creativity, but it also brings negative beliefs of the type "I'm not doing a good job" or "I 'm not good enough".

And well, going to the printer and drinking coffee are definitely different types of tasks!

So instead, try this:

1.   Put a timer and write freely without reading what you write. No editing. Just thinking and writing as if you were speaking. You can start with 10 minutes, and then increase it to 25 min (pomodoro-style).

2.   Then it's time to edit. Change what you don't like, delete words, and insert citations if you remember. Still, you're editing, not reading, so keep the reading time for another moment.

Free writing is a skill that will boost your creativity and your writing. And guess what, you can then always use AI to gather some help to edit the text that you wrote and has your ideas.

There are a lot of mini strategies and habits that you can stack before and after this strategy. But the best is that you experiment with what works for you.

P.S. I wrote this text following this strategy. I did not use chatGPT. But I passed my text through Grammarly to catch any mistakes. Be sure you do that too!

Andrea E. A. Stephens
Editor-in-Chief, Trends in Ecology and Evolution

When a paper hits the desk of an editor, they need to decide whether to send it for peer review or to "desk reject" it. The question that the editor is asking themselves is *"Is this paper likely to be successful at my journal, given (1) my journal's scope and (2) my journal's requirement for a conceptual advance?"*.

Every journal has its own niche. Your chance of success greatly increases when you find a journal whose niche matches your paper.

Let's take the two editorial criteria separately. First, scope. This is relatively straightforward. Every journal publishes their "Aims and Scope" somewhere on their website. Read these in detail and ask yourself whether your paper aligns with these.

Second, conceptual advance, often referred to as "novelty". A major conceptual advance will be something that clarifies a complex problem, opens up a new line of inquiry or changes our way of thinking. A less major conceptual advance might be to demonstrate that a phenomenon is broader than initially thought. The level of conceptual advance can be hard to assess, can be subjective and, in general, is more important for a broad scope journal than a journal with a narrower scope.

While some high-profile journals have a requirement for a groundbreaking conceptual advance, most work does not meet this bar. As such there are many journals that are happy with papers that where the advance is more limited or even for purely confirmatory work

(sometimes called "solid science" journals). These journals are well-respected and good work published in them is well-read and cited.

As long as your science is done well, there is a home for your work. Desk rejects are not a judgement on your scientific ability, they simply indicate that the editor feels that your paper doesn't match their journal's niche. Identifying a journal where your science will be read will mean that your science will have the impact you are hoping for.

Nianxun Xi
Hainan University, China

Early-career researchers often face numerous challenges and difficulties during the process of publishing their papers, with one of the primary obstacles being flawed or even incorrect experimental design (e.g., pseudo-replication). They need to invest effort into their writing to enhance the chances of acceptance. Drawing from my personal experience, I advise early-career researchers to carefully consider the following points to improve their likelihood of acceptance.

In my opinion, the introduction is the most important part of a paper. It is essential to articulate a comprehensive theoretical or conceptual framework that helps readers quickly grasp the logic behind your research. If necessary, these frameworks can be illustrated using figures. This framework naturally leads to the scientific questions or hypotheses to be addressed (I recommend clearly stating testable hypotheses). It is important to note that your experimental design and data analysis should be based on the questions or hypotheses being investigated.

In the introduction, methods, and other sections of your paper, it is crucial to avoid content that is unrelated to your research. You may collect a large amount of data, but not all of it is relevant to the study.

Including irrelevant information can disrupt the logical flow of your paper, making it difficult for readers to grasp the research theme and thereby diminishing the quality of your work and increasing the likelihood of rejection. Therefore, avoid presenting data that, although collected, does not contribute to your research objectives. Keep in mind the Occam's Razor: "Entities must not be multiplied beyond necessity".

In the discussion section, it is important to highlight the novelty of your research and illustrate how your findings enhance our understanding of nature. However, you should be cautious not to overinterpret your results, as this can detract from the credibility of your research. I believe that a high-quality paper should embody logic, conciseness, and novelty.

Claudia Ratti
Physicist

By far the most overlooked—and powerful—principle in scientific writing is this: clarity is kindness. Many early-career researchers believe they must sound "academic" to be taken seriously. They bury their ideas under jargon, long sentences, and passive constructions. But science is hard enough. Don't make your reader fight your writing.

Clarity isn't about dumbing things down. It's about being generous to your reader. A well-structured, plainly written paper gives your work the best possible chance to be understood, appreciated, and cited.

One strategy I recommend is to write your paper as if you're explaining it to a smart colleague from another field. Imagine someone who knows research, but not your niche. This keeps your writing honest. It also forces you to define terms, explain assumptions, and focus on the core message—something even reviewers appreciate more than you think.

A simple but powerful tool is **reverse outlining**. After you draft a section, write down the main point of each paragraph in a few words. If the points don't follow a logical flow, your reader will get lost. Rearranging or rewriting at this stage often improves clarity dramatically.

Another common barrier—especially for non-native English speakers

—is confidence. Many assume they must become "great writers" before they can publish. That's not true. You don't need to be a stylist. You need to be clear, honest, and well-organized. Don't wait for perfect English—aim for clear thinking, then get help polishing the language.

Finally, remember that scientific writing is a craft. No one is born knowing how to do it. The more you write, revise, and read good papers, the better you'll become. Be patient with yourself. A rejection isn't a verdict on your intelligence—it's a signal for revision. Treat it like an experiment. What can you learn?

Mariano A. Rodriguez-Cabal
Ecologist

If you are just beginning your career in science, chances are you were drawn to the field by a fascination with the natural world, not a passion for writing. And yet, writing will determine the trajectory of your scientific life more than any single method, analysis, or field site. It is how you will win grants, build collaborations, share discoveries, and perhaps most importantly clarify your own thinking.

Despite its centrality to our work, writing is often treated as an afterthought in scientific training. Students are taught how to design experiments and collect data, but rarely how to craft a compelling manuscript or proposal. As the Editor-in-Chief of a journal that reads hundreds of submissions each year, I can tell you: this gap in training is visible, but it is also doable.

Let me be clear, writing is not a gift possessed by a lucky few. It is a skill. Like statistical modeling or species identification, it improves with practice, feedback, and repetition. The most successful scientists I know treat writing as part of the scientific process itself. They use it not only to communicate ideas, but to discover them.

The act of writing whether a manuscript, a grant, a conference talk, or even a figure caption, forces you to make your ideas precise. Many researchers believe they understand a topic until they try to explain it in words. That is when the uncertainty surfaces. As you search for the right words, you begin to see gaps in your logic, unanswered questions in your data, or weaknesses in your framing. Writing makes your thinking visible. In that way, it is both diagnostic and generative.

This is why I advise early career researchers to write early and often. Do not wait until your data are "perfect" or your argument fully formed. Start drafting now! even if all you have is a hypothesis or a rough methods section. You might be surprised by how quickly the act of writing, especially the final paragraph of the introduction, where you articulate the study's relevance, hypotheses/questions, and the methods section can help clarify your thinking and guide your data analysis.

Scientific writing is not a performance, it's a process. You don't need to be eloquent; you need to be clear. You don't need to sound impressive; you need to be understood. And above all, you need to keep writing. The sooner you do, the faster you will grow, not just as a writer, but as a scientist.

Kulbhushansingh Suryawanshi
Conservationist, India

"What is the value of an academic publication if I am not interested in an academic job after my degree?" asked one of my brightest students, who was also among the most passionate about doing conservation at the grassroots. She wanted to bring about change on the ground, here and now.

The slow pace of academic publishing and what she called its 'excessive focus on the details' were frustrating for her. The dull and drab language and rigid structure of a scientific paper were uninspiring. Yet,

I knew that her conservation research work needed to be published in an academic journal for it to reach its full potential in achieving conservation on the ground.

Over the course of the semester, we began unpacking some of the myths and misconceptions around scientific publishing, which helped lower a few of the initial barriers. It's not just academic publishing that is slow; all carefully produced publications take time. Only click- bait social media content is generated quickly. In class, we read The Ghost of Competition Past by Connell (1980) to understand that academic writing does not have to be dull and boring. It can be creative while still being accurate and precise — and that should be an aspiration for every scientist, young or old.

We also spoke about the peer-review process and the sanctity of this system. The conversation meandered from "why some reviewers are mean" to "can peer review actually damage the quality of a paper,"

but in the end, we agreed that the process plays a vital role in verifying the claims of the proponent and, in doing so, lends credibility to the paper. In many countries, research papers have been used as evidence in courtrooms when fighting environmental litigations. This was the most convincing argument for the 'passionate conservationists' in my class to start taking research publishing more seriously. We called it 'arming the activists' to fight the just fight for environmental conservation.

The students who engaged deeply in the discussion also observed that knowing their paper would face strict peer review forced them to consider all possible interpretations of their data and to resist jumping to the most exciting conclusion. That kind of careful and systematic thinking was essential — we needed to channel that passion for conservation into research in order to make a lasting impact on the ground. A good research paper about a conservation problem meant it was more likely that others would build upon that work. It also meant you were more likely to get funding for the conservation project that emerged from the research. Research papers improved the credibility of conservation projects.

Young conservationists in the global south often struggle to have their voices heard, and my students agreed that a well-argued research paper gives them credibility to engage with policymakers and bureaucrats — it earns them a seat at the table.

We concluded that academic publishing can indeed be a powerful tool in a conservationist's toolbox. Even when it feels like they are publishing the most obvious results, the larger world depends on them to produce rigorous evidence-based solutions to pressing conservation problems.

Reference: *Connell, J. H. (1980). Diversity and the coevolution of competitors, or the ghost of competition past. Oikos, 131–138.*

Gabriel Rabinovich
Immunologist, tumor biologist, and glycobiologist

Writing has played a key role in my career as a researcher. I firmly believe that one does not simply write up results sequentially; instead, one tells a story. I've always enjoyed storytelling. After completing the experiments, I try to craft a narrative that appeals to an interdisciplinary audience, not just to specialists in my field. My goal is always for people from other disciplines to be able to understand my papers. Writing in a linear, step-by-step way can feel monotonous and dull. I try to construct a compelling story. Sometimes, it's not the question I asked at the beginning, but the one that emerges from all the results taken together. This approach has had a big impact on how my papers are received. When editors read something that tackles a meaningful question in the field, they're more likely to be positive with it. That's why I put a lot of effort into the cover letters. They're the first thing editors see, and I structure them point-by-point with clear logic to show that the paper addresses an open question in the literature.

One piece of advice I wish I'd received earlier is to write up results as you go. Write them as soon as you've finished the experiments; it's much harder if you wait. Try to make it a habit to build the story as you go along, and by the time you've completed some good experiments, you may already have the core of a paper. I usually start by preparing the figures, their legends, materials and methods, and the results section. Only afterward do I write the introduction and discussion. I never write the paper sequentially. I also recommend keeping a research journal, and having students do the same—documenting what worked and what didn't, day by day.

To begin writing, I always try to have the figures in front of me. You need a clear idea of your results. Once you've sketched out the data, you can write the methods. Then you decide on the narrative for the introduction—how you're going to frame and "sell" the story.

The hardest part for me is always writing the discussion. You have to be fair with the literature. The worst thing you can do is ignore or downplay another published study. Even if you think it has technical issues, you have to engage with it because it's out there. That's why discussions are so demanding for me—I want to reconcile the literature, to fairly consider both the work that supports my findings and the work that points in the opposite direction. It's not about promoting your own perspective, but about being as honest and inclusive as possible. A critical part of the discussion should also be a paragraph on the study's limitations. Every study has them. Acknowledging them shows integrity.

If you're submitting to a top-tier immunology journal, you need to show that you're answering a compelling, unresolved question in the field. Of course, reviewers may point out technical issues, but I try not to allow conceptual gaps in the questions I ask. As an editor, I often see that the main problem is a lack of clarity about the central question—just a series of results with no clear hypothesis. The biggest mistakes I see researchers make aren't about grammar or English. They're about conceptual clarity. People often don't really know what they want to convey. They get lost in the details and fail to see the big picture. They struggle to ask broad, integrative questions. That's why I always ask: what's the story you want to tell? What do you want to demonstrate? If you're not constantly clear on your experimental goals, you lose track of the specific objectives and, ultimately, of the overall purpose. That's when things fall apart.

Another problem that I see is that some researchers are so specialized that they struggle to connect even closely related fields. That's why I think scientists who teach are better—they're forced to take a broader view of their work. I also love attending outsider conferences—it helps avoid intellectual inbreeding.

I don't choose journals the conventional way. I rarely follow impact factors, which don't always reflect a journal's true importance. I prefer to submit to interdisciplinary journals so that people from different fields can engage with the work. If we've answered a truly disruptive question, I aim for the highest-profile journals—even if that means years of revisions and additional experiments. If it's a major advance but not fully disruptive, I target other multidisciplinary journals. If it's just one piece of a broader puzzle, then I look to more specialized journals, ideally from science societies. The key is to be self-critical and realistic, so no one wastes time. Anyway rejections happen to all, while the first ones can feel devastating, they get easier with time.[*]

[*] *This text is an extract from an interview made by M. Nuñez.*

OTHER RESOURCES

Here I mention some of the books that have been influential to me and that have helped me be a better scientific writer.

* **Bird by Bird: Some Instructions on Writing and Life –** *Anne Lamott.*

While not strictly about scientific writing, this classic book offers essential lessons on overcoming writer's block, embracing the writing process, and improving clarity in communication. Also is where I learned about the "shitty first draft" idea.

* **Science Research Writing: /or Native and Non-Native Speakers of English** – *Hilary Glasman-Deal.*

A step-by-step guide tailored for scientists who need to write papers in English, with a structured approach to constructing clear and effective manuscripts.

* **Writing Science in Plain English** – *Anne E. Greene.*

Focuses on simplifying scientific language and making research accessible, emphasizing clarity and engagement.

*   **How to Write and Illustrate a Scientific Paper** – *Björn Gustavii.*

Covers both the writing and visual presentation aspects of scientific papers, offering practical advice for crafting clear figures and tables.

*   **How to Write a Scientific Paper: An Academic Self-Help Guide for PhD Students** – *Jari Saramäki.*

A practical guide designed to help PhD students write and publish their first scientific papers.

*   **An Editor's Guide to Writing and Publishing Science** – *Michael Hochberg.*

Provides insight from an editor's perspective on what makes a paper successful, covering everything from writing to navigating peer review and publishing. This is one of my favorite books with a lot of really valuable information.

*   **The Elements of Style** – *William Strunk Jr. & E.B. White.*

A concise, classic guide to writing clearly and effectively, widely used by researchers of all levels and students.

GLOSSARY

- A -

APC (Article Processing Charge): A fee charged to authors to make their article openly accessible. This is common in many open-access journals and varies widely depending on the journal's impact and publisher.

- C -

Clarity: In scientific writing, clarity means expressing ideas in a way that leaves no room for confusion or misinterpretation. A clear manuscript is easy to follow, with a logical flow and straightforward language.

Copy editor: A professional responsible for correcting grammar, punctuation, and style in a manuscript before publication. In scientific publishing, they do not typically change content but ensure consistency and readability.

- D -

Desk reject: When a manuscript is rejected by the editor without being sent for peer review. This can happen if the paper is out of scope, poorly written, or lacks novelty.

Diamond open access: A publishing model where articles are freely available to all readers without any fees for authors or institutions. These journals are often funded by universities, research institutions, or scholarly societies, making them fully open-access without financial barriers.

- F -

For-Profit publisher: A commercial publisher that generates revenue primarily through subscriptions, article processing charges (APCs), or other publishing-related fees. Examples include Elsevier, Springer Nature, and Wiley. These publishers often operate large portfolios of journals across various disciplines.

- G -

Ghost writer: In science, a ghostwriter is someone who writes or significantly edits a paper without being credited as an author. This sometimes happens in fields like medicine, where busy professionals (e.g., physicians) may rely on a writer to help publish research based on their data or clinical insights.

Gold Open Access: A model in which an article is made freely available immediately upon publication, usually requiring the author (or their institution/funder) to pay an APC. This ensures unrestricted access to the research but can be costly for authors.

Google Scholar: A fully free search engine that indexes scholarly articles, conference papers, and other academic content across disciplines. It is widely used but does not have the same level of indexing control as Web of Science or Scopus.

Green Open Access: A model where authors can archive a version of their paper (often a preprint or accepted manuscript) in a publicly accessible repository, such as an institutional archive or preprint server. Unlike Gold Open Access, the published version often remains behind a paywall. Some journals impose embargo periods before allowing Green OA deposits.

- H -

Hybrid Open Access: A publishing model where traditional subscription-based journals allow authors to pay an APC to make their individual article open access. This creates a mix of paywalled and freely available articles within the same journal, leading to concerns about double-dipping (charging both authors and subscribers).

H-Index: A metric used to measure a researcher's productivity and citation impact. A scholar has an H-index of n if they have published n papers that have each been cited at least n times. For example, an H-index of 10 means the researcher has 10 papers, each cited at least 10 times. While it helps assess research influence, it has limitations— it favors older researchers with more publications and does not account for author position in multi-author papers.

- I -

Impact factor (IF): A metric used to measure the average number of citations a journal's articles receive in a given year. It is calculated by dividing the number of citations in a specific year by the total number of articles published in the previous two years. Higher impact factors generally indicate greater influence, but the metric has limitations, including field-dependent variability and susceptibility to manipulation. It is frequently used to assess journal prestige but does not necessarily reflect the quality of individual papers.

- N -

Not-for-Profit publisher: A publishing organization that reinvests revenue into scholarly activities rather than distributing profits to shareholders. Many society journals, university presses, and open-access initiatives fall into this category. Examples include PLOS, eLife, and journals published by professional societies such as the Ecological Society of America.

- O -

Open Access Publishers: Publishers that specialize in open-access journals, where all articles are freely available to readers. Some operate under ethical, non-profit models (e.g., PLOS), while others are for-profit and charge high APCs. The quality of open-access publishers varies widely, so authors must assess credibility before submitting.

- P -

Paywall: A barrier that prevents access to an article or journal without a subscription or payment. Many traditional academic publishers keep articles behind paywalls unless an open-access fee is paid.

Plagiarism: Plagiarism is the use of someone else's words, ideas, or data without proper citation. This includes copying, paraphrasing without attribution, or presenting others' work as your own. Even unintentional plagiarism is a serious ethical violation that can lead to manuscript rejection, retractions, or professional consequences.

Peer Review: The process by which a submitted manuscript is evaluated by experts in the field before publication. It ensures the quality, validity, and originality of research.

PI: Principal Investigator. Is the person responsible for a given lab. Can be a PhD., a Masters or Postdoc advisor

Predatory journals: Journals that exploit the open-access model by charging authors publication fees without providing proper peer review, editorial oversight, or indexing in reputable databases. These journals often solicit submissions aggressively, promise unrealistically fast publication times, and lack transparency about editorial policies. Publishing in a predatory journal can damage a researcher's reputation and limit the impact of their work.

Precision: In scientific writing, precision means avoiding ambiguity and providing information at the necessary level of detail. A precise statement is exact, specific, and unambiguous, ensuring that the reader gets the most accurate interpretation possible. However, precision does not always guarantee clarity—a sentence can be highly detailed but still difficult to understand. Conversely, a sentence can be clear but not precise if it is too vague or general. Good scientific writing balances both clarity and precision—ensuring that information is not only correct and specific but also easily understood.

Preprint: A version of a scientific manuscript that is shared publicly before undergoing formal peer review. Preprints are typically posted on specialized servers (e.g., arXiv, bioRxiv, medRxiv) to allow rapid dissemination of research findings. Some journals accept submissions that were previously posted as preprints, while others do not.

Published: A manuscript is considered published when it has been formally released by a journal, whether in print or online. In some cases, preprints and accepted manuscripts online may also be considered published.

- R -

Reviewer: A scientist who evaluates a manuscript during the peer-review process, providing feedback and recommendations to the journal editor. Reviewers assess originality, methodology, clarity, and validity.

- S -

Salami science: is the practice of splitting a single substantial study into multiple smaller papers to increase publication counts rather than presenting the full research in one comprehensive article. While dividing a large dataset can be valid when addressing distinct questions, unjustified slicing leads to redundant publications and fragmented literature. This practice can make findings harder to interpret, inflate publication records, and is often frowned upon by journals, reviewers and CV evaluators.

Scopus: A large database of academic literature covering journals, conference proceedings, and patents. It is often used for citation tracking and evaluating research impact.

Self-Plagiarism: Self-plagiarism occurs when an author reuses their own previously published work without disclosure. This includes copying text, figures, or data without citation, misleading readers about the novelty of the text. While some repetition is unavoidable, transparency is required to avoid ethical issues. I have mixedfeelings about this, since there are not so many ways you can write some parts of your paper (e.g., study site), but it is not acceptable to use previous text.

Senior Editor: An experienced editor who handles manuscripts, assigns reviewers, and makes decisions on whether a paper should be sent for peer review or rejected. They typically work under the Editor-in-Chief.

Society Published: A journal that is published by a scientific society rather than a commercial publisher. These journals often have strong reputations in their fields and may offer lower APCs for society members.

Subject Editor: A journal editor responsible for handling submissions within a specific area of expertise. They assign reviewers and make recommendations on publication decisions.

- T -

Traditional publishing: The standard publishing model where journals are subscription-based, meaning readers (or their institutions/libraries) must pay for access to articles. Authors do not typically pay to publish, but their work may be behind a paywall.

- W -

Web of science: A widely used citation database that indexes high-quality journals across disciplines. It is often used for research evaluation and journal impact metrics.

ACKNOWLEDGEMENTS

So many people to thank here. First the people that has helped me with the book. They are Romina Dimarco, Dan Simberloff, Jaime Moyano, and Laura Meyerson. Their help has improved the book tremendously and made it much better. A very special thank you to Walter Policelli, who created the figures, designed the cover, and did the layout. His creativity and design sense gave the book its unique visual style. Thank you!

Also the many authors that have contributed their texts to this book. I loved to read their perspective and I'm sure you'll love them too.

And last but not least all the people that have helped me become a better writer and editor. Working on the British Ecological Society has been especially important for me to understand more about publishing. Also all the people that had read my many early drafts have made me a better and/or more confident writer. They are many but include Nate Sanders, Dan Simberloff, Romina Dimarco, Anibal Pauchard, Aimee Classen, Mariano Rodriguez-Cabal, Laura Meyerson, Estela Rafaelle, and many many others in this more than 2 decades of publishing.

Thanks to all!!

ABOUT THE AUTHOR

Dr. Martin A. Nuñez is an Argentinean scientist and professor at the *University of Houston*. He has published over 200 scientific papers, reviewed hundreds more, and handled more than 2,500 manuscripts as an editor. He has served as editor, senior editor, and editor-in-chief of international scientific journals.

Keep in touch with Martin

x.com/Martin_A_Nunez

bsky.app/profile/martin-nunez.bsky.social

linkedin.com/in/martin-andres-nunez

instagram.com/martin.a.nunez/